Grounding for the Control of EMI

By Hugh W. Denny

Electromagnetic Compatibility Division
Electronics and Computer Systems Laboratory
Engineering Experiment Station
Georgia Institute of Technology

emf-emi control, Inc.
6193 Finchingfield Rd.
Gainesville, VA 22065
Telephone: 703-347-0030   Fax: 703-347-5813

© Copyright 1983
First Edition
Sixth Printing 1989

All rights reserved. This book, or any parts thereof, may not be reproduced in any form without the written permission of the publisher.

Library of Congress Catalog Card No. 82-063066
ISBN# 0-932263-17-8

Printed in the United States of America

Acknowledgement

Words of sincere gratitude and thanks are extended to several individuals who encouraged this effort or were otherwise instrumental in formulating some of the concepts contained in this book. In particular, I wish to express my appreciation to Margie, my wife, who continually encouraged my efforts when my enthusiasm waned. Next, words of thanks are due to Mr. Robert Goldblum of R and B Enterprises who periodically checked on my progress; to Mr. Donald R.J. White of Don White Consultants, Inc., who exhibited extreme patience with my plodding progress; and to the reviewers who made many helpful suggestions. Lastly, I wish to thank my colleagues in the Electromagnetic Compatibility Division, Electronics and Computer Systems Laboratory, Engineering Experiment Station, Georgia Institute of Technology, who assisted in the formulation of many of the concepts in this book; in particular, the many spirited discussions with Jimmy A. Woody which were responsible for honing down many of the finer points presented.

Also I wish to thank those at Don White Consultants, Inc. for the logistics in publishing this book. This includes editing by Edward R. Price; typing by Colleen S. White and Marie Price; drafting by William Horn and the cover design by Gilbert Fitzpatrick.

Foreword

It is with great pleasure on behalf of Don White Consultants, Inc. (DWCI) that I release *Grounding For The Control of EMI*. This is another of our new handbooks on EMC and related topics to be published.

All of our handbooks are prepared by recognized experts in their field. Several are now in preparation. DWCI's role is to provide the technical guidance, editing, logistics, financing, publishing and promotion. These books will provide a major contribution to the EMC and related technologies for years to come.

Regarding this handbook, *Grounding For The Control of EMI*, it fills an existing void. This book has been prepared for engineers/designers as well as technicians who are engaged in related technology and its application in the electronics industry. This handbook also has been written in terms of the advanced state-of-the-art and carefully illustrated in such a manner that it can be used in tutorial and seminar courses, as well as at undergraduate levels of instruction. Therefore, it provides an invaluable design guide, and an adjunct to existing literature, dealing with this important and dynamic substantive area. The author, Hugh W. Denny invites your comments. Similarly DWCI welcomes correspondence from the many readers who may wish to comment on any aspect of this book.

January 1983
Gainesville, Virginia USA   Donald R.J. White

Handbooks Published by EEC

(1) White, Donald R.J., *Electrical Filters—Synthesis, Design & Applications,* 1980.
(2) White, Donald R.J., Volume 1, *Electrical Noise and EMI Specifications,* 1971.
(3) White, Donald R.J., Volume 2, *Electromagnetic Interference Test Methods and Procedures,* 1980.
(4) White, Donald R.J., Volume 3, *Electromagnetic Interference Control Methods & Techniques,* 1973.
(5) White, Donald R.J., Volume 4, *Electromagnetic Interference Test Instrumentation Systems,* 1980.
(6) Duff, Dr. William G. and White, Donald R.J., Volume 5, Electromagnetic *Interference Prediction & Analysis Techniques,* 1972.
(7) Hill, James S. and White, Donald R.J., Volume 6, *Electromagnetic Interference Specifications, Standards & Regulations,* 1975.
(8) White, Donald R.J., *A Handbook on Electromagnetic Shielding Materials and Performance,* 1980.
(9) Duff, Dr. William G., *A Handbook on Mobile Communications,* 1980.
(10) White, Donald R.J., *EMI Control Methodology & Procedures,* 1982.
(11) White, Donald R.J., *EMI Control in the Design of Printed Circuit Boards and Backplanes,* 1982. (Also available in French.)
(12) Jansky, Donald M., *Spectrum Management Techniques,* 1977.
(13) Herman, John R., *Electromagnetic Ambients and Man-Made Noise,* 1979.
(14) Hart, William C. and Malone, Edgar W., *Lightning and Lightning Protection,* 1979.
(15) Kaiser, Dr. Bernhard E., *EMI Control in Aerospace Systems,* 1979.
(16) Feher, Dr. Kamilo, *Digital Modulation Techniques in an Interference Environment,* 1977.
(17) Gard, Michael F., *Electromagnetic Interference Control in Medical Electronics,* 1979.
(18) Carstensen, Russell V., *EMI Control in Boats and Ships,* 1979.
(19) Georgopoulos, Dr. Chris J., *Fiber Optics and Optical Isolators,* 1982.
(20) Mardiguian, Michel, *How to Control Electrical Noise,* 1983.
(21) Denny, Hugh W., *Grounding for Control of EMI,* 1983.
(22) Ghose, Rabindra N., *EMP Environment and System Hardness Design,* 1983.
(23) Mardiguian, Michel, *Interference Control in Computers and Microprocessor-Based Equipment,* 1984.
(24) *EMC Technology 1982 Anthology*
(25) Mardiguian, Michel, *Electrostatic Discharge—Understand, Simulate and Fix ESD Problems*
(26) White, Donald R.J., *Shielding Design Methodology and Procedures*

Notice

All of the books listed above are available for purchase from *emf-emi control, Inc.*, 6193 Finchingfield Road, Gainesville, VA 22065 USA. Telephone: 703-347-0030, Fax: 703-347-5813

Preface

Grounding is considered by some to be an art. In support of this view is the fact that many signal grounding systems are totally unstructured and the reason why some systems work, or do not work, is not clear. As a result, there is a continual search for a *set of rules* that can be used for the design and installation of a grounding system. Unfortunately, many of the rules are not in agreement. For example, one rule says that ground conductors shall not exceed 0.05λ while another says that 0.15λ is appropriate; yet another ignores length considerations altogether. As a result, curious combinations of the various rules produce grounding systems that seemingly do not work. For rules to work at all, they must be based on clear scientific principles. A grounding rule which is valid for shock or fire protection at power frequencies may not be valid for the control of Electromagnetic Interference (EMI) in a strong Ultra High Frequency (UHF) environment. The purpose of this volume on *Grounding for the Control of EMI* is to present the scientific principles governing the performance of grounding networks in all frequency regions.

The goals of this volume are to establish a proper perspective of the functions of various grounding systems and to help the reader correctly analyze and design grounding networks most appropriate for a particular system. Toward these ends, the various rules of grounding networks are examined, basic principles of operation are described and a set of preferred design practices are presented. Specifically, Chapter 1 briefly examines the electromagnetic environment within which grounding networks must perform and identifies the multiple roles that networks serve. Chapter 2 describes the lumped circuit and transmission line nature of grounding networks and relates it to interference coupling. Means for controlling coupling are explored in Chapter 3. With these principles in mind, Chapter 4 lays out pragmatic approaches for realizing effective ground systems. Considerations of grounding in terms of personnel safety and fault protection are covered in Chapter 5. Chapters 6 and 7 integrate the collective principles into applications for electronic systems and equipment, respectively. Chapter 8 examines the mechanical and electrical behavior aspects of the ground network interconnects, i.e., the bonds, while Chapter 9 sets forth some elemental ground system tests and maintenance recommendations.

Table of Contents
Grounding for the Control of EMI

	Page No.
ACKNOWLEDGEMENT	i
FOREWARD	ii
BOOKS PUBLISHED BY EEC	iii
PREFACE	v
TABLE OF CONTENTS	vii
ILLUSTRATIONS & TABLES	xi

Chapter 1 Introduction to Grounding

1.1	THE ENVIRONMENT	1.1
1.2	WHY GROUND?	1.3
1.3	REFERENCES	1.6

Chapter 2 Ground Circuit Behavior

2.1	INTERFERENCE	2.1
2.2	GROUND CONDUCTOR PROPERTIES	2.5
2.3	FREQUENCY EFFECTS	2.14
2.4	COMMON IMPEDANCE COUPLING	2.19
2.5	INTERFERENCE COUPLING	2.21
2.6	REFERENCES	2.24

Chapter 3 Control of Unwanted Coupling

3.1	GROUND PLANE IMPEDANCE CONTROL	3.1

Table of Contents

3.2	OPENING GROUND LOOPS		3.5
	3.2.1	Single-Point Grounding	3.5
	3.2.2	Common-Mode Rejection	3.5
	3.2.3	Frequency Translation	3.7
	3.2.4	Optical Isolation	3.7
	3.2.5	Frequency-Selective Grounding	3.8
3.3	REFERENCES		3.10

Chapter 4　Ground Network Configurations

4.1	FLOATING GROUND		4.2
4.2	SINGLE-POINT GROUND		4.3
4.3	MULTIPLE-POINT GROUND		4.8
	4.3.1	Requirements	4.9
	4.3.2	Applications	4.11
4.4	RECOMMENDED APPROACH		4.14
	4.4.1	Low-Frequency Network	4.15
	4.4.2	High-Frequency Network	4.15
	4.4.3	Selection of the Dividing Frequency	4.16
4.5	REFERENCES		4.18

Chapter 5　Grounding for Fault Protection

5.1	ELECTRIC SHOCK	5.1
5.2	FAULT PROTECTION OBJECTIVES	5.3
5.3	FAULT PROTECTION DESIGN	5.3
5.4	REFERENCES	5.8

Chapter 6　System Grounding

6.1	THE DESIGNER'S DILEMMA		6.3
6.2	BASIC SYSTEM GROUNDING REQUIREMENTS		6.6
	6.2.1	Isolated System	6.6
	6.2.2	Clustered System	6.8
	6.2.3	Distributed System	6.12
	6.2.4	Multiple Distributed Systems	6.13
	6.2.5	Central System With Extensions	6.14

Table of Contents

6.3	SINGLE-POINT GROUNDING	6.16
6.4	CONCLUSIONS	6.20
6.5	REFERENCES	6.20

Chapter 7 Equipment Grounding

7.1	CIRCUIT RETURN COUPLING	7.2
7.2	LOW-FREQUENCY CIRCUIT GROUNDING	7.3
7.3	POWER SUPPLY FILTER GROUNDING	7.8
7.4	DIGITAL CIRCUITS	7.10
7.5	INSTRUMENTATION GROUNDING	7.15
	7.5.1 Grounded Transducers	7.15
	7.5.2 Ungrounded Transducers	7.19
	7.5.3 Transducer Amplifiers	7.20
7.6	CONSTRUCTION GUIDELINES FOR LOW FREQUENCY EQUIPMENT	7.21
7.7	HIGH-FREQUENCY CIRCUIT GROUNDING	7.22
7.8	HYBRID CIRCUITS	7.24
7.9	SHIELD GROUNDING	7.25
7.10	REFERENCES	7.28

Chapter 8 Bonding

8.1	EFFECTS OF POOR BONDS	8.1
8.2	BOND RESISTANCE	8.3
8.3	DIRECT BONDS	8.4
	8.3.1 Welding	8.4
	8.3.2 Brazing	8.5
	8.3.3 Soft Solder	8.5
	8.3.4 Bolts	8.5
	8.3.5 Conductive Adhesive	8.5
8.4	INDIRECT BONDS	8.6
8.5	BOND CORROSION	8.9
	8.5.1 Chemical Basis of Corrosion	8.9

Table of Contents

	8.5.2 Bond Protection Code	8.11
	8.5.3 Jumper Fasteners	8.13
8.6	WORKMANSHIP	8.15
8.7	EQUIPMENT BONDING PRACTICES	8.18
8.8	SUMMARY OF BONDING PRINCIPLES	8.24
8.9	REFERENCES	8.25

Chapter 9 Ground Systems Tests and Maintenance

9.1	TEST PROCEDURES	9.1
	9.1.1 Bond Resistance	9.2
	9.1.2 Ground System Noise Current	9.3
	9.1.3 Differential Noise Voltage	9.4
9.2	MAINTENANCE	9.7
	9.2.1 Inspection	9.7
	9.2.2 Bus System Measurements	9.7
	9.2.3 Corrective Action	9.12
9.3	REFERENCES	9.15

Index

Illustrations & Tables

Figure No.	Title	Page No.

Chapter 1 Introduction to Grounding

1.1	Grounding Scenario	1.2
1.2	The Multiple Functions of Grounds	1.4
1.3	Grounding for Fault Clearance	1.5
1.4	The Varied Functions of Grounding	1.6

Chapter 2 Ground Circuit Behavior

2.1	Idealized Energy Transfer Loop	2.2
2.2	Energy Transfer Loop with Noise Sources in Ground System	2.2
2.3	Equivalent Circuit of Non-Ideal Energy Transfer Loop	2.2
2.4	Practical Combinations of Source-Load Pairs	2.4
2.5	Ground Strap Inductance as a Function of Width	2.8
2.6	Ground Strap Inductance as a Function of Strap Thickness	2.8
2.7	Ground Strap Inductance as a Function of Length	2.9
2.8	Relative Inductive Reactance Versus Length-to-Width Ratio of Flat Straps	2.10
2.9	Ground Strap Resonance Effects	2.11
2.10	The Equivalent Circuit of a Ground Cable Parallel to a Ground Plane	2.14
2.11	Idealized Equipment Grounding	2.16
2.12	Typical Impedance vs Frequency Behavior of a Grounding Conductor	2.17
2.13	Photograph of the Swept Frequency Behavior of a Grounding Strap	2.17
2.14	Coupling Between Circuits Caused by Common Return Path Impedance	2.19
2.15	Conductive Coupling of Extraneous Noise into Equipment Interconnecting Cables	2.20
2.16	Summary of Coupling Modes	2.22

Chapter 3 Control of Unwanted Coupling

3.1	Multiple Grounding of Cable Trays	3.2

Illustrations

3.2	Structural Bonding	3.2
3.3	Example of Large Ground Busses	3.3
3.4	Application of Large Ground Busses to Control Interference	3.3
3.5	Use of Differential Line Receivers for Common-Mode Rejection	3.6
3.6	Balanced Operation With Balanced-to-Unbalanced Transformers	3.6
3.7	Common-Mode Chokes	3.7
3.8	Use of Optical Isolation to Combat Common-Mode Noise	3.8
3.9	Capacitive Grounding	3.9
3.10	Inductive Grounding	3.9

Chapter 4 Ground Network Configurations

4.1	Floating Signal Ground	4.2
4.2	Single-Point Signal Ground	4.4
4.3	Single-Point Ground Bus System Using Separate Risers	4.4
4.4	Single-Point Ground Bus System Using a Common Bus	4.5
4.5	Zonal Grounding	4.6
4.6	Use of Single-Point Ground Configuration to Minimize Effect of Facility Ground Currents	4.7
4.7	Multiple-Point Ground Configuration	4.8
4.8	Use of Structural Steel in Multiple-Point Grounding	4.9
4.9	Multiple Equipment Grounding	4.11
4.10	Common Multiple-Point Ground	4.12

Chapter 5 Grounding for Fault Protection

5.1	Single Phase 115/230 Volt AC Power Ground Connections	5.4
5.2	Three Phase 120/208 Volt AC Power System Ground Connections	5.5
5.3	Properly Wired AC Distribution Circuits for Minimum Ground Noise	5.6
5.4	Noise Problems Resulting from Improper Wiring	5.7

Chapter 6 System Grounding

6.1	Elements of a Facility Ground System	6.4
6.2	Minimum Grounding Requirements for an AC Powered Isolated System	6.7
6.3	Minimum Grounding Requirements for Battery Powered Isolated System	6.8
6.4	A Clustered System	6.9
6.5	Basic Grounding of a Clustered System	6.10
6.6	Grounding of a Common Battery Clustered System	6.10
6.7	Signal Grounding Schemes for Clustered Systems	6.11

Illustrations

6.8	Central-With-Extensions System	6.14
6.9	Conducted Noise Threat	6.17
6.10	Preferred Grounding of Low Frequency Equipment in Noisy Environments	6.17
6.11	Alternative No. 1 for Single-Point Grounding	6.18
6.12	Alternative No. 2 for Single-Point Grounding	6.19

Chapter 7 Equipment Grounding

7.1	Circuit Return Coupling	7.2
7.2	Ground Plane Induced Feedback	7.3
7.3	Interstage Coupling Caused by Improper Grounding	7.4
7.4	Recommended Grounding Scheme for PC Board Mounted Chain Amplifiers	7.5
7.5	Grounding of Potentially Incompatible Circuits Sharing a Common Circuit Board	7.6
7.6	The Star Ground	7.6
7.7	Power Supply Hum Caused by Ground Return Coupling	7.8
7.8	Grounding of Power Supply Filter Capacitors to Minimize Ground Noise	7.9
7.9	Single-Point Grounding of Logic Systems	7.11
7.10	A Single-Point Grounded System	7.12
7.11	Grounding in Low Frequency Equipment	7.13
7.12	An Alternate Single Point Logic Ground System	7.14
7.13	Grounding Practices for Single-Ended Amplifiers	7.16
7.14	Grounding Practices for Differential Amplifiers	7.16
7.15	Method of Grounding Bridge Transducers	7.17
7.16	Use of Isolated Differential Amplifier with Balanced Bridge Transducer	7.18
7.17	Recommended Grounding Practices for Floating Transducers	7.19
7.18	Establishment of Shield Continuity Between High Frequency Equipment	7.23
7.19	Grounding Practices in Equipment Containing Both High Frequency and Low Frequency Circuits	7.24
7.20	Grounding of Overall Cable Shields to Connectors and Penetrated Walls	7.26
7.21	Grounding of Overall Cable Shields to Terminal Strips	7.27

Chapter 8 Bonding

8.1	Effects of Poor Bonding on the Performance of a Power Line Filter	8.2
8.2	Inductive Reactance of Wire and Strap Bond Jumpers	8.7
8.3	True Equivalent Circuit of an Indirect Bonding System	8.8
8.4	Techniques for Protecting Bonds Between	

Illustrations & Tables

	Dissimilar Metals	8.11
8.5	Order of Assembly for Bolted Connection	8.17
8.6	Bonding of Subassemblies to Equipment Chassis	8.19
8.7	Bonding of Equipment to Mounting Surface	8.20
8.8	Typical Method of Bonding Equipment Flanges to Frame or Rack	8.21
8.9	Bonding of Rack-Mounted Equipment Employing Dagger Pins	8.21
8.10	Recommended Practices for Effective Bonding in Cabinets	8.22
8.11	Bonding of Connector to Mounting Surface	8.23

Chapter 9 Ground Systems Tests and Maintenance

9.1	Bond Resistance Measurement	9.3
9.2	Test Setup for Stray Current Measurements	9.4
9.3	Oscilloscope Connections for Measuring Voltage Levels on Ground Systems	9.5
9.4	Typical Results Provided by Differential Noise Voltage Test	9.6
9.5	Method for Determining the Existence of Improper Neutral Ground Connections	9.9
9.6	Typical Bond Resistance and Stray Current Measurement Locations in an Electronic Facility	9.10
9.7	Determination of Single-Point Ground System Compromises with the Use of Induced Currents	9.13

Tables

2.1	Properties of Annealed Copper Wire	2.6
2.2	Resistance Properties of Grounding Straps	2.6
2.3	Inductance of 15 cm Rectangular Straps	2.7
2.4	Inductance of 1.0 mm Thick Straps	2.7
2.5	Inductance of Standard Size Cable	2.7
2.6	Metal Ground Plane Impedance in Ohms/Square	2.12
2.7	Impedance of Straight Circular Copper Wires	2.13
4.1	Various Suggested Dividing Frequencies Between Single Point and Multiple Point Grounding	4.17
5.1	Summary of the Effects of Shock	5.2
8.1	Galvanic Series of Common Metals and Alloys	8.10
8.2	Compatible Groups of Common Metals	8.11
8.3	Bond Protection Requirements	8.13
8.4	Metal Connections for Aluminum and Copper Jumpers	8.14

CHAPTER 1

Introduction to Grounding

Grounding is essential to the protection against electrical shock and a vital element of lightning protection. Its role in the protection of equipment and systems against Electromagnetic Interference (EMI) is not thoroughly understood. This misunderstanding frequently results from a failure to consider all factors, such as environment, which determine how well a particular grounding system or network performs. This chapter addresses *environment* and reviews the various reasons *why grounding is necessary*.

1.1 The Environment

Large numbers of diverse circuits* are usually involved in the makeup of an electronic system. Whether distances between individual circuits are large or small, the entire system must function as an integral unit. Each circuit must perform its intended function and supply an output to a designated load, in an interference-free manner, in the presence of extraneous signals. Grounding of circuits is an essential ingredient of this process.

Typically, a system must operate in an environment containing many potentially incompatible (error-producing or damage-threatening) voltages and currents as illustrated in Fig. 1.1. For example, within a facility are power sources (operating at frequencies specified as dc, 50 Hz, 60 Hz, and 400 Hz); very low frequency signals from monitors, indicators and other specialized devices; and audio frequency voltages and currents associated with voice communications and control systems. In the higher frequency region of the spectrum there are Radio Frequency (RF) signals, ranging from Very Low Frequencies (VLF) to microwaves used for broadcast communications, surveillance, tracking and other functions. Extending from audio

---

* A circuit is considered to be any collection of passive and active elements combined to perform a specific function, i.e., attenuate, amplify, rectify, detect, filter, or otherwise alter a waveform. A grounding path is also considered to be a circuit or part of a circuit.

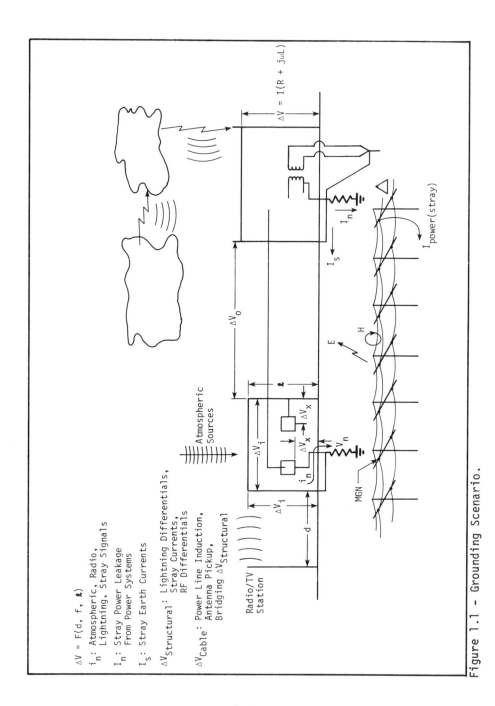

Figure 1.1 - Grounding Scenario.

frequencies into the RF region are the broadband data and communications systems, both analog and digital. Lightning discharges and stray earth currents further contribute to the noise enviroment.[1] These various signals, falling in overlapping frequency ranges and representing a wide range of amplitudes, pose a definite threat of interference (and possible damage) to devices unless careful measures are taken to minimize their coupling into susceptible circuits. Equipment, system and facility grounding is important in minimizing interference from sources internal or external to a system.

1.2 Why Ground?

Historically, grounding requirements arose from the need to provide protection from lightning strokes and industrially-generated static electricity. Structures and electrical equipment were connected to earth, i.e., grounded, to provide necessary conduction paths for lightning and static discharges. As utility power transmission systems developed, grounding to earth was found to be necessary for personnel and equipment safety. All major components of a transmission system such as generating stations, substations, and distribution elements had to be earth grounded to provide a path back to the generator for the fault currents in case of line trouble.

With the development of electronics, metal became the preferable choice for structural and enclosure construction because it provided fire protection, mechanical strength, and EMI control. Therefore, many grounding problems are perhaps related to the presence of metal. The ready availability of something which can be used as a *ground* tends to obscure the real reason why a ground is needed. Metal associated with electrical circuitry poses a possible shock hazard which frequently can lead to arguments about *earthing* the metal for an electronic ground when perhaps the parts should not be metallic in the first place.[2] (Double insulated tools, which eliminate exposed, energizable metal surfaces, do not need to be grounded).

Multiple electronic circuits and equipment often must share common metallic paths which, in turn, may also serve as power returns, lightning discharge paths, an integral portion of an electromagnetic shield, etc., as illustrated in Fig. 1.2. Numerous currents from various sources may be present in the common impedance of a path which frequently leads to undesired EMI coupling. Effective grounding is the realization of an appropriate reference network serving multiple roles without producing EMI between user circuits and equipment. In essence, the purpose of grounding is to electrically interconnect conductive or charged objects in order to minimize the potential differences between them. Functionally, grounds typically provide a:[3]

- low resistance connection with earth to provide a

Sec. 1.2   Why Ground?

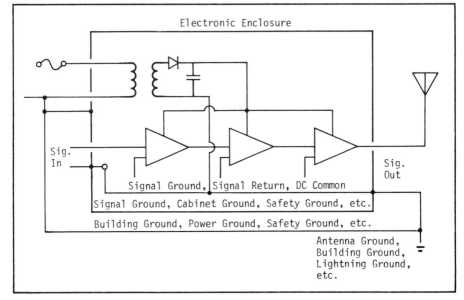

Figure 1.2 - The Multiple Functions of Grounds.

    fault return path between the fault and the source to lessen the voltage hazard until fuses blow or breakers trip (Fig. 1.3).

- low resistance path between electrical/electronic equipment and nearby metallic objects to minimize personnel danger in the event of an electrical fault within the equipment (Fig. 1.4a).

- preferential path between the point of impringement of a lightning stroke on an exposed object and the earth (Fig. 1.4b).

- path for bleeding-off static charge before the potential becomes high enough to produce a spark or an arc (Fig. 1.4c).

- common reference plane of low relative impedance between electronic devices, circuits, and systems.

- reference plane for long wave antenna systems.

    Many electronic grounding systems simultaneously involve two or more of these functions. For example, one interconnected metallic

Sec. 1.2   Why Ground?

Figure 1.3 - Grounding for Fault Clearance.

system or network may serve both safety and EMI control functions
and also perform as part of an antenna system. Frequently, such
multiple roles are in conflict either in terms of operational requirements or in terms of techniques of implementation. A basic intent
of the remaining chapters of this book is to foster an understanding
of the electromagnetic properties of grounding networks so that designers can appropriately configure networks needed in particular
facilities, systems, equipment, and circuits and have high degrees
of confidence in their performance.

Sec. 1.2   Why Ground?

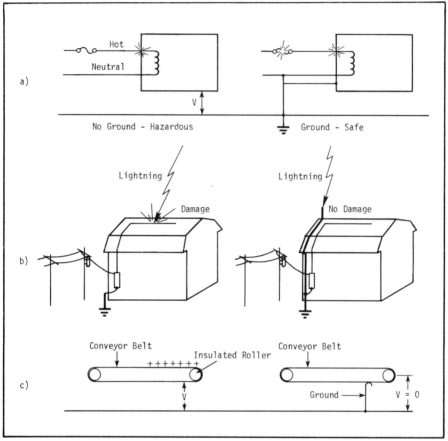

Figure 1.4 - The Varied Functions of Grounding.

1.3 References

1. Herman, J.R., *Electromagnetic Ambients and Man-Made Noise*, Volume III, EMC Encyclopedia Series, Don White Consultants, Inc. Gainesville, Virginia, 1979.

2. Herring, T.H., *Large System Grounding*, FAA/GIT Workshop on the Grounding of Electronic Systems, Atlanta, Ga., Report No. FAA-RD-74-174, March 1974, pp. 109-120.

3. Military Specification MIL-B-5087B, *Bonding, Electrical and Lightning Protection, for Aerospace Systems*, 15 October 1964.

CHAPTER 2

Ground Circuit Behavior

Ground circuit conductors possess resistance, inductance, and capacitance relative to other objects. Considering these properties, a simplified projection of the frequency-dependent behavior of a grounding path can be made. From this characteristic behavior, it can be shown that ground conductors must be short relative to the wavelength of potentially interfering signals. Principles of coupling between conductors establish a basis for ground loop avoidance techniques presented in later chapters.

2.1 Interference

Interference is any extraneous electrical or electromagnetic disturbance that (1) tends to disturb the reception of desired signals or (2) produces undesirable responses in a circuit or system. Interference can be produced by both natural and man-made sources either external or internal to the circuit. The correct operation of complex electronic equipment and facilities is inherently dependent upon the frequencies and amplitudes of both the signals utilized in the system and the potential interference emissions that are present. If the frequency of an undesired signal is within the operating frequency range of a circuit, the circuit may respond to the undesired signal (it may even happen out-of-band). The severity of the interference is a function of the amplitude of the undesired signal relative to that of the desired signal at the point of detection.

An ideal signal system is a simple signal generator-lead pair as shown in Fig. 2.1.[1] With no extraneous voltages present within the loop, this simple pair is free of interference. Consider, however, what happens when the current return path is not ideal and sources of noise are present as shown in Fig. 2.2. Unless noise voltages V_{N1} and V_{N2} are identical, a voltage difference will exist between the low side of the generator (Node 1) and the low side of the load (Node 2). This voltage difference effectively appears in the signal transfer loop in series with the signal generator and produces noise currents in the load (see Fig. 2.3).

Sec. 2.1   Interference

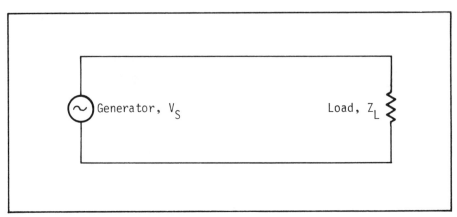
Figure 2.1 - Idealized Energy Transfer Loop.

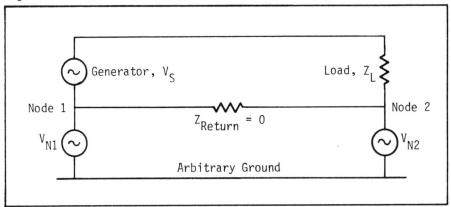
Figure 2.2 - Energy Transfer Loop with Noise Sources in Ground System.

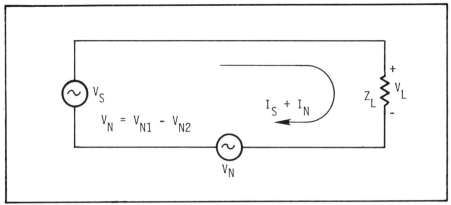
Figure 2.3 - Equivalent Circuit of Non-Ideal Energy Transfer Loop.

Practical electronic circuits typically are collections of several source-load combinations such as are shown in Fig. 2.4. Within such circuits, numerous sources and loads are interconnected and the use of individual return paths for each source-load pair becomes impractical. It is often more realistic to establish a common ground or reference plane which serves as the return path for several signals. For example, the control of undesired network responses, particularly in high gain or high frequency circuits, often requires establishment of a common signal reference to which related groupings of components, circuits, and networks can be connected. Ideally, this common reference connection offers zero impedance paths to all signals for which it serves as a reference. The several signal currents within the individual networks can then return to their respective sources without creating unwanted coupling between the separate circuits. Unfortunately, grounding circuits are not ideal and thus do not provide zero impedance paths for power, signal, or any other currents.

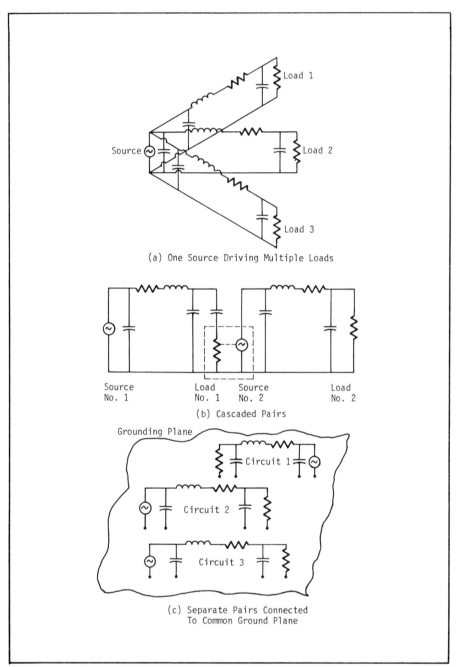

Figure 2.4 - Practical Combinations of Source-Load Pairs.

2.2 Ground Conductor Properties

Every element (conductor) of a grounding system, whether it be for power grounding, signal grounding, or lightning protection, has properties of resistance, capacitance, and inductance. Shields and drain wires of signal cables, the *green wire* power ground, lightning down conductors, transformer vault buses, structural steel members -- all conductors have these properties. The resistance property is exhibited by all metals. The resistance of a ground path conductor is a function of the material, its length, and its cross-sectional area. The capacitance associated with a ground conductor is determined by its geometric shape, its proximity to other conductors, and the nature of the intervening dielectric. The inductance is a function of its size, geometry, length, and, to a limited extent, the relative permeability of the metal.

Table 2.1 lists the physical dimensions and the resistance properties of standard American gauge copper wires; Table 2.2 lists the per unit length resistance, R/ℓ, of representative grounding/bonding straps as calculated from Eq. (2.1):

$$R/\ell = \frac{\rho}{A} \qquad (2.1)$$

where ρ is the resistivity of the material and A is the cross-sectional area of the strap in consistent units.

Assessment of the capacitance associated with a ground conductor requires knowledge of its surroundings. The inductance, however, may be calculated for rectangular conductors from:[2]

$$L = 0.002\ell \left[2.303 \log \frac{2\ell}{b+c} + 0.5 + 0.2235 \frac{b+c}{\ell} \right] \mu H \qquad (2.2)$$

where b is the width of the strap, c is its thickness, and ℓ its length in centimeters. For circular wires, Eq. (2.2) reduces to:

$$L = 0.002\ell \left[2.303 \log \frac{4\ell}{d} - 0.75 \right] \mu H \qquad (2.3)$$

where ℓ is the length and d is the diameter of the wire in centimeters.

Tables 2.3 and 2.4 give the inductance values of typical rectangular straps of selected dimensions. Table 2.5 lists the inductances of various lengths of standard sized wires. Figures 2.5, 2.6 and 2.7 illustrate the relative behavior of the inductance of rectangular straps as a function of width, thickness, and length, respectively. Observe that the inductance does decrease with increasing width and increases with increasing length. Thus, a recommendation

Sec. 2.2 Ground Conductor Properties

Table 2.1 - Properties of Annealed Copper Wire

| AWG | Diameter | | Cross-Sectional Area | | Resistance in Ohms | |
No	mils	mm	cmil	mm^2	per 1000 ft	per km
4/0	460.0	11.7	211,600	107.2	0.049	0.161
3/0	409.6	10.4	167,800	85.0	0.062	0.203
2/0	364.8	9.3	133,100	67.4	0.078	0.256
1/0	324.9	8.3	105,500	53.4	0.098	0.322
1	289.3	7.3	83,690	42.4	0.124	0.407
2	257.6	6.5	66,370	33.6	0.156	0.512
4	204.3	5.2	41,740	21.1	0.248	0.814
6	162.0	4.1	26,250	13.3	0.395	1.296
8	128.5	3.3	16,510	8.4	0.628	2,060
10	101.9	2.6	10,380	5.3	0.999	3.278
12	80.8	2.1	6,530	3.3	1.588	5.210
14	64.1	1.6	4,107	2.1	2.525	8.284
16	50.8	1.3	2,583	1.3	4.016	13.176
18	40.3	1.0	1,624	0.8	6.385	20.948
20	31.9	0.8	1,022	0.5	10.150	33.300

Table 2.2 - Resistance Properties of Grounding Straps

| Strap Size | Unit Length Resistance in $\mu\Omega$/cm | | |
(cm)	Copper	Aluminum	Steel
0.05 x 0.5	69.2	107.6	388.9
0.05 x 1.0	34.6	53.8	194.5
0.05 x 2.0	17.3	26.9	97.2
0.05 x 5.0	6.9	10.8	38.8
0.05 x 10.0	3.5	5.4	19.7
0.1 x 0.5	34.6	53.8	194.5
0.1 x 1.0	17.3	26.9	97.2
0.1 x 2.0	8.7	13.5	48.9
0.1 x 5.0	3.5	5.4	19.7
0.1 x 10.0	1.7	2.7	9.6

Table 2.3 - Inductance of 15 cm Rectangular Straps

Width, b, in cm	Thickness, c, in mm	L in µH
1.0	0.2	0.112
1.0	1.0	0.099
1.0	2.0	0.085
2.5	0.2	0.088
2.5	1.0	0.081
2.5	2.0	0.074
5.0	0.2	0.070
5.0	1.0	0.066
5.0	2.0	0.062

Table 2.4 - Inductance of 1.0 mm Thick Straps

	Inductance in µH		
Width (cm)	ℓ = 15 cm	30 cm	100 cm
1	0.115	0.270	1.14
2.5	0.090	0.220	0.970
5.0	0.070	0.180	0.836

Table 2.5 - Inductance of Standard Size Cable

	Inductance in µH		
AWG Gauge	15 cm	30 cm	100 cm
4/0	0.096	0.233	1.017
1/0	0.106	0.253	1.086
2	0.113	0.268	1.135
4	0.120	0.282	1.179
6	0.127	0.296	1.227
10	0.141	0.323	1.318
14	0.155	0.352	1.415

Ground Conductor Properties

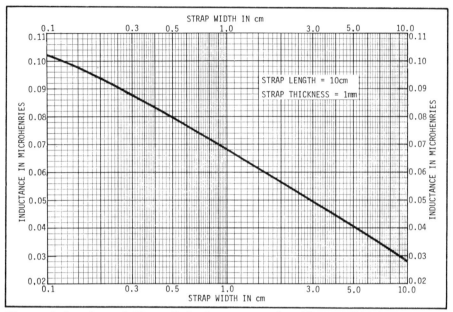

Figure 2.5 - Ground Strap Inductance as a Function of Width.

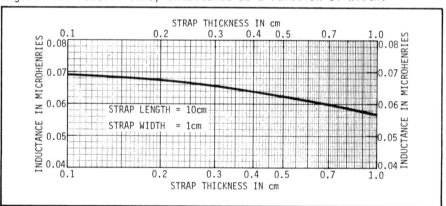

Figure 2.6 - Ground Strap Inductance as a Function of Strap Thickness.

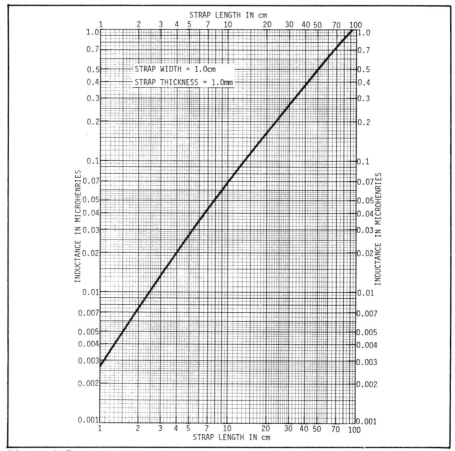

Figure 2.7 - Ground Strap Inductance as a Function of Length.

frequently encountered is that of restricting the length-to-width ratio to 5:1 for grounding and bonding straps. The behavior of the strap reactance relative to a straight round wire, as a function of length-to-width ratio is illustrated by Fig. 2.8. Note that a 5:1 length-to-width ratio indicates a reactance of about 45% of that of a straight round wire.

Figure 2.9 shows the resonant frequencies associated with a 50 cm by 30 cm cabinet, chassis, or circuit board separated from a ground plane with various thickness spaces of air and teflon. It is grounded to the ground plane with 1 cm x 1 mm ground strap of varying lengths. Note that practical and reasonable lengths of less than 30 cm can easily produce resonant behavior in the HF region. These resonances can be particularly important in EMI control in strong

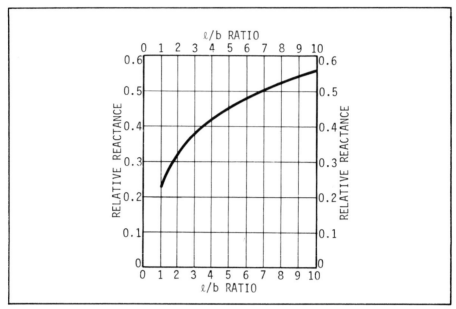

Figure 2.8 - Relative Inductive Reactance Versus Length-to-Width Ratio of Flat Straps.[3]

radiated environments by actually enhancing pickup rather than suppressing it.

Even large sheet ground planes do not offer zero impedances although they will offer three to four orders of magnitude of impedance less than a single small wire. Typical ground plane impedances in ohms/square are given in Table 2.6. For direct comparison with straight copper wires, see Table 2.7.

Sec. 2.2 Ground Conductor Properties

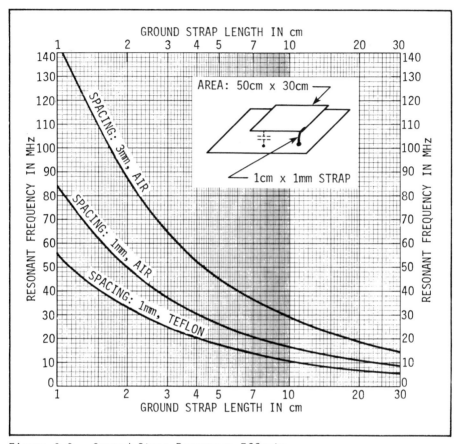

Figure 2.9 - Ground Strap Resonance Effects.

Sec. 2.2   Ground Conductor Properties

Table 2.6 - Metal Ground Plane Impedance in Ohms/Squares[4]

Freq.	COPPER, COND-1, PERM-1						STEEL, COND-17, PERM-200					
	t=.03	t=.1	t=.3	t=1	t=3	t=10	t=.03	t=.1	t=.3	t=1	t=3	t=10
10Hz	574μ	172μ	57.4μ	17.2μ	5.74μ	1.75μ	3.38m	1.01m	338μ	101μ	38.5μ	40.3μ
20Hz	574μ	172μ	57.4μ	17.2μ	5.75μ	1.83μ	3.38m	1.01m	338μ	102μ	49.5μ	56.6μ
30Hz	574μ	172μ	57.4μ	17.2μ	5.75μ	1.95μ	3.38m	1.01m	338μ	103μ	62.3μ	69.3μ
50Hz	574μ	172μ	57.4μ	17.2μ	5.76μ	2.30μ	3.38m	1.01m	338μ	106μ	86.2μ	89.6μ
70Hz	574μ	172μ	57.4μ	17.2μ	5.78μ	2.71μ	3.38m	1.01m	338μ	110μ	105μ	106μ
100Hz	574μ	172μ	57.4μ	17.2μ	5.82μ	3.35μ	3.38m	1.01m	338μ	118μ	127μ	126μ
200Hz	574μ	172μ	57.4μ	17.2μ	6.04μ	5.16μ	3.38m	1.01m	340μ	157μ	179μ	179μ
300Hz	574μ	172μ	57.4μ	17.2μ	6.38μ	6.43μ	3.38m	1.01m	342μ	199μ	219μ	219μ
500Hz	574μ	172μ	57.4μ	17.3μ	7.36μ	8.27μ	3.38m	1.01m	350μ	275μ	283μ	283μ
700Hz	574μ	172μ	57.4μ	17.3μ	8.55μ	9.77μ	3.38m	1.01m	362μ	335μ	335μ	335μ
1kHz	574μ	172μ	57.4μ	17.5μ	10.4μ	11.6μ	3.38m	1.01m	385μ	403μ	400μ	400μ
2kHz	574μ	172μ	57.5μ	18.3μ	16.1μ	16.5μ	3.38m	1.02m	495μ	566μ	566μ	566μ
3kHz	574μ	172μ	57.5μ	19.5μ	20.3μ	20.2μ	3.38m	1.03m	623μ	693μ	694μ	694μ
5kHz	574μ	172μ	57.6μ	23.0μ	26.2μ	26.1μ	3.38m	1.06m	862μ	896μ	896μ	896μ
7kHz	574μ	172μ	57.8μ	27.1μ	30.9μ	30.9μ	3.38m	1.10m	1.05m	1.06m	1.06m	1.06m
10kHz	574μ	172μ	58.2μ	33.5μ	36.9μ	36.9μ	3.38m	1.18m	1.27m	1.26m	1.26m	1.26m
20kHz	574μ	172μ	60.4μ	51.6μ	52.2μ	52.2μ	3.40m	1.57m	1.79m	1.79m	1.79m	1.79m
30kHz	574μ	172μ	63.8μ	64.3μ	63.9μ	63.9μ	3.42m	1.99m	2.19m	2.19m	2.19m	2.19m
50kHz	574μ	173μ	73.6μ	82.7μ	82.6μ	82.6μ	3.50m	2.75m	2.83m	2.83m	2.83m	2.83m
70kHz	574μ	173μ	85.5μ	97.7μ	97.7μ	97.7μ	3.62m	3.35m	3.35m	3.35m	3.35m	3.35m
100kHz	574μ	175μ	104μ	116μ	116μ	116μ	3.85m	4.03m	4.00m	4.00m	4.00m	4.00m
200kHz	575μ	183μ	161μ	165μ	165μ	165μ	4.95m	5.66m	5.66m	5.66m	5.66m	5.66m
300kHz	575μ	195μ	203μ	202μ	202μ	202μ	6.23m	6.93m	6.94m	6.94m	6.94m	6.94m
500kHz	576μ	230μ	262μ	261μ	261μ	261μ	8.62m	8.96m	8.96m	8.96m	8.96m	8.96m
700kHz	578μ	271μ	309μ	309μ	309μ	309μ	10.5m	10.6m	10.6m	10.6m	10.6m	10.6m
1MHz	582μ	335μ	369μ	369μ	369μ	369μ	12.7m	12.6m	12.6m	12.6m	12.6m	12.6m
2MHz	604μ	516μ	522μ	522μ	522μ	522μ	17.9m	17.9m	17.9m	17.9m	17.9m	17.9m
3MHz	638μ	643μ	639μ	639μ	639μ	639μ	21.9m	21.9m	21.9m	21.9m	21.9m	21.9m
5MHz	736μ	827μ	826μ	826μ	826μ	826μ	28.3m	28.3m	28.3m	28.3m	28.3m	28.3m
7MHz	855μ	977μ	977μ	977μ	977μ	977μ	33.5m	33.5m	33.5m	33.5m	33.5m	33.5m
10MHz	1.04m	1.16m	1.16m	1.16m	1.16m	1.16m	40.0m	40.0m	40.0m	40.0m	40.0m	40.0m
20MHz	1.61m	1.65m	1.65m	1.65m	1.65m	1.65m	56.6m	56.6m	56.6m	56.6m	56.6m	56.6m
30MHz	2.03m	2.02m	2.02m	2.02m	2.02m	2.02m	69.4m	69.4m	69.4m	69.4m	69.4m	69.4m
50MHz	2.62m	2.61m	2.61m	2.61m	2.61m	2.61m	89.6m	89.6m	89.6m	89.6m	89.6m	89.6m
70MHz	3.09m	3.09m	3.09m	3.09m	3.09m	3.09m	106m	106m	106m	106m	106m	106m
100MHz	3.69m	3.69m	3.69m	3.69m	3.69m	3.69m	126m	126m	126m	126m	126m	126m
200MHz	5.22m	5.22m	5.22m	5.22m	5.22m	5.22m	179m	179m	179m	179m	179m	179m
300MHz	6.39m	6.39m	6.39m	6.39m	6.39m	6.39m	219m	219m	219m	219m	219m	219m
500MHz	8.26m	8.26m	8.26m	8.26m	8.26m	8.26m	283m	283m	283m	283m	283m	283m
700MHz	9.77m	9.77m	9.77m	9.77m	9.77m	9.77m	335m	335m	335m	335m	335m	335m
1GHz	11.6m	11.6m	11.6m	11.6m	11.6m	11.6m	400m	400m	400m	400m	400m	400m
2GHz	16.5m	16.5m	16.5m	16.5m	16.5m	16.5m	566m	566m	566m	566m	566m	566m
3GHz	20.2m	20.2m	20.2m	20.2m	20.2m	20.2m	694m	694m	694m	694m	694m	694m
5GHz	26.1m	26.1m	26.1m	26.1m	26.1m	26.1m	896m	896m	896m	896m	896m	896m
7GHz	30.9m	30.9m	30.9m	30.9m	30.9m	30.9m	1.06Ω	1.06Ω	1.06Ω	1.06Ω	1.06Ω	1.06Ω
10GHz	36.9m	36.9m	36.9m	36.9m	36.9m	36.9m	1.26Ω	1.26Ω	1.26Ω	1.26Ω	1.26Ω	1.26Ω

* t is in units of mm
μ = microhms
m = milliohms
Ω = ohms

NOTE: Do not use table at frequencies in MHz above $5/\ell_m$ since the separation distance in meters, ℓ_m, of two grounded equipments will exceed 0.05λ where error becomes significant.

Sec. 2.2   Ground Conductor Properties

Table 2.7 - Impedance of Straight Circular Copper Wires[4]

	AWG#=2, D=6.54mm				AWG#=10, D=2.59mm				AWG#=22, D=.64mm			
FREQ.	ℓ=1cm	ℓ=10cm	ℓ=1m	ℓ=10m	ℓ=1cm	ℓ=10cm	ℓ=1m	ℓ=10m	ℓ=1cm	ℓ=10cm	ℓ=1m	ℓ=10m
10Hz	5.13μ	51.4μ	517μ	5.22m	32.7μ	327μ	3.28m	32.8m	529μ	5.29m	52.9m	529m
20Hz	5.14μ	52.0μ	532μ	5.50m	32.7μ	328μ	3.28m	32.8m	529μ	5.29m	53.0m	530m
30Hz	5.15μ	52.8μ	555μ	5.94m	32.8μ	328μ	3.28m	32.9m	529μ	5.30m	53.0m	530m
50Hz	5.20μ	55.5μ	624μ	7.16m	32.8μ	329μ	3.30m	33.2m	530μ	5.30m	53.0m	530m
70Hz	5.27μ	59.3μ	715μ	8.68m	32.8μ	330μ	3.33m	33.7m	530μ	5.30m	53.0m	530m
100Hz	5.41μ	66.7μ	877μ	11.2m	32.9μ	332μ	3.38m	34.6m	530μ	5.30m	53.0m	530m
200Hz	6.20μ	99.5μ	1.51m	20.6m	33.2μ	345μ	3.67m	39.6m	530μ	5.30m	53.0m	530m
300Hz	7.32μ	137μ	2.19m	30.4m	33.7μ	365μ	4.11m	46.9m	530μ	5.30m	53.0m	531m
500Hz	10.1μ	219μ	3.59m	50.3m	35.3μ	425μ	5.28m	64.8m	530μ	5.31m	53.2m	533m
700Hz	13.2μ	303μ	5.01m	70.2m	37.1μ	500μ	6.66m	84.8m	530μ	5.32m	53.4m	537m
1kHz	18.1μ	429μ	7.14m	100m	42.2μ	632μ	8.91m	116m	531μ	5.34m	53.9m	545m
2kHz	35.2μ	855μ	14.2m	200m	62.5μ	1.13m	16.8m	225m	536μ	5.48m	56.6m	589m
3kHz	52.5μ	1.28m	21.3m	300m	86.3μ	1.65m	25.0m	336m	545μ	5.71m	60.9m	656m
5kHz	87.3μ	2.13m	35.6m	500m	137μ	2.72m	41.5m	559m	571μ	6.39m	72.9m	835m
7kHz	122μ	2.98m	49.8m	700m	189μ	3.79m	58.1m	783m	609μ	7.28m	87.9m	1.04Ω
10kHz	174μ	4.26m	71.2m	1.00Ω	268μ	5.41m	82.9m	1.11Ω	681μ	8.89m	113m	1.39Ω
20kHz	348μ	8.53m	142m	2.00Ω	533μ	10.8m	165m	2.23Ω	1.00m	15.2m	207m	2.63Ω
30kHz	523μ	12.8m	213m	3.00Ω	799μ	16.2m	248m	3.35Ω	1.39m	22.0m	305m	3.91Ω
50kHz	871μ	21.3m	356m	5.00Ω	1.33m	27.0m	414m	5.58Ω	2.20m	36.1m	504m	6.48Ω
70kHz	1.22m	29.8m	498m	7.00Ω	1.86m	37.8m	580m	7.82Ω	3.04m	50.2m	704m	9.06Ω
100kHz	1.74m	42.6m	712m	10.0Ω	2.66m	54.0m	828m	11.1Ω	4.31m	71.6m	1.00Ω	12.9Ω
200kHz	3.48m	85.3m	1.42Ω	20.0Ω	5.32m	108m	1.65Ω	22.3Ω	8.59m	142m	2.00Ω	25.8Ω
300kHz	5.23m	128m	2.13Ω	30.0Ω	7.98m	162m	2.48Ω	33.5Ω	12.8m	214m	3.01Ω	38.7Ω
500kHz	8.71m	213m	3.56Ω	50.0Ω	13.3m	270m	4.14Ω	55.8Ω	21.4m	357m	5.01Ω	64.6Ω
700kHz	12.2m	298m	4.98Ω	70.0Ω	18.6m	378m	5.80Ω	78.2Ω	30.0m	500m	7.02Ω	90.4Ω
1MHz	17.4m	426m	7.12Ω	100Ω	26.6m	540m	8.28Ω	111Ω	42.8m	714m	10.0Ω	129Ω
2MHz	34.8m	853m	14.2Ω	200Ω	53.2m	1.08Ω	16.5Ω	223Ω	85.7m	1.42Ω	20.0Ω	258Ω
3MHz	52.3m	1.28Ω	21.3Ω	300Ω	79.8m	1.62Ω	24.8Ω	335Ω	128m	2.14Ω	30.1Ω	387Ω
5MHz	87.1m	2.13Ω	35.6Ω	500Ω	133m	2.70Ω	41.4Ω	558Ω	214m	3.57Ω	50.1Ω	646Ω
7MHz	122m	2.98Ω	49.8Ω	700Ω	186m	3.78Ω	58.0Ω	782Ω	300m	5.00Ω	70.2Ω	904Ω
10MHz	174m	4.26Ω	71.2Ω	1.00kΩ	266m	5.40Ω	82.8Ω	1.11kΩ	428m	7.14Ω	100Ω	1.29kΩ
20MHz	348m	8.53Ω	142Ω	2.00kΩ	532m	10.8Ω	165Ω	2.23kΩ	857m	14.2Ω	200Ω	2.58kΩ
30MHz	523m	12.8Ω	213Ω	3.00kΩ	798m	16.2Ω	248Ω	3.35kΩ	1.28Ω	21.4Ω	301Ω	3.87kΩ
50MHz	871m	21.3Ω	356Ω	5.00kΩ	1.33Ω	27.0Ω	414Ω	5.58kΩ	2.14Ω	35.7Ω	501Ω	6.46kΩ
70MHz	1.22Ω	29.8Ω	498Ω	7.00kΩ	1.86Ω	37.8Ω	580Ω	7.82kΩ	3.00Ω	50.0Ω	702Ω	9.04kΩ
100MHz	1.74Ω	42.6Ω	712Ω	10.0kΩ	2.66Ω	54.0Ω	828Ω	11.1kΩ	4.28Ω	71.4Ω	1.00kΩ	12.9kΩ
200MHz	3.48Ω	85.3Ω	1.42kΩ	20.0kΩ	5.32Ω	108Ω	1.65kΩ	22.3kΩ	8.57Ω	142Ω	2.00kΩ	25.8kΩ
300MHz	5.23Ω	128Ω	2.13kΩ	30.0kΩ	7.98Ω	162Ω	2.48kΩ	33.5kΩ	12.8Ω	214Ω	3.01kΩ	38.7kΩ
500MHz	8.71Ω	213Ω	3.56kΩ	50.0kΩ	13.3Ω	270Ω	4.14kΩ	55.8kΩ	21.4Ω	357Ω	5.01kΩ	64.6kΩ
700MHz	12.2Ω	298Ω	4.98kΩ	70.0kΩ	18.6Ω	378Ω	5.80kΩ	78.2kΩ	30.0Ω	500Ω	7.02kΩ	90.4kΩ
1GHz	17.4Ω	426Ω	7.12kΩ		26.6Ω	540Ω	8.28kΩ		42.8Ω	714Ω	10.0kΩ	

* AWG = American Wire Gage
D = wire diameter in mm
ℓ = wire length in cm or m
μ = microhms
m = milliohms
Ω = ohms

☐ Non-Valid Region for which ℓ ≥ λ/4

2.3 Frequency Effects

A situation commonly encountered is that of a ground cable (power or signal) running along in the proximity of a ground plane. A representative circuit of this *simple* ground path is illustrated by Fig. 2.10. The effects of the resistive elements of the *circuit* will predominate at very low frequencies. The relative influence of the reactive elements will increase at increasing frequencies. At some frequency, the magnitude of the inductive reactance ($j\omega L$) equals the magnitude of the capacitive reactance ($1/j\omega C$) and the circuit becomes resonant. The frequency of the primary (or first) resonance can be determined from:

$$f = \frac{1}{2\pi\sqrt{LC}} \qquad (2.4)$$

where L is the total cable inductance and C is the net capacitance between the cable and the ground plane. At resonance, the impedance presented by the grounding path will either be high or low, depending on whether it is parallel or series resonant, respectively. At parallel resonance, the impedance seen looking into one end of the cable will be much higher than expected from $R + j\omega L$. (For good conductors, e.g., copper and aluminum, $R \ll \omega L$; thus $j\omega L$ generally provides an accurate estimate of the impedance of a ground conductor at frequencies above a few hundred Hertz). At parallel resonance:

$$Z_p = Q\omega L \qquad (2.5)$$

where, Q, the quality factor, is defined as:

$$Q = \frac{\omega L}{R_{(ac)}} \qquad (2.6)$$

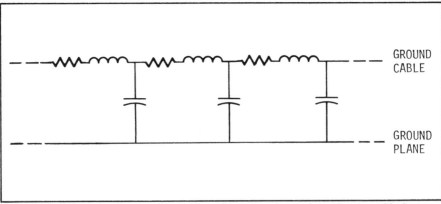

Figure 2.10 - The Equivalent Circuit of a Ground Cable Parallel to a Ground Plane.

Sec. 2.3   Frequency Effects

where $R_{(ac)}$ is the cable resistance at the frequency of resonance.

Then:

$$Z_p = Q\omega L = \frac{\omega L}{R_{(ac)}} \times \omega L = \frac{\omega^2 L^2}{R_{(ac)}} \qquad (2.7)$$

Above the primary resonance, subsequent resonances (both parallel and series) will occur between the various possible combinations of inductances and capacitances (including parasitics) in the path.

Series resonances in the grounding *circuit* will also occur between the inductances of line (wire) segments and one or more of the shunt capacitances. The impedance of a series resonant path is:

$$Z_s = \frac{\omega L}{Q} \qquad (2.8)$$

therefore,

$$Z_s = \omega L \frac{\omega L}{R_{(ac)}} = R_{(ac)} \qquad (2.9)$$

The series resonant impedance is thus determined by, and is equal to, the series ac resistance of the particular inductance and capacitance in resonance. (At the higher ordered resonances, where the resonant frequency is established by line (wire) segments and not the total path, the series impedance of the path to ground may be less than predicted from a consideration of the entire ground conductor length).

An understanding of the high frequency behavior of a grounding conductor is simplified by viewing it as a transmission line. If the ground path is considered uniform along its run, well known techniques[5] can be used to describe the voltages and currents along the line as a function of time and distance. If the resistance elements in Fig. 2.11 are small relative to the inductances and capacitances, the grounding path has a characteristic impedance, Z_0, equal to $\sqrt{L/C}$ where L and C are the per unit length values of inductance and capacitance.

The situation illustrated in Fig. 2.11 is of particular interest in equipment grounding. The input impedance of the grounding path, i.e., the impedance to ground seen by the equipment case, is:[5]

$$Z_{in} = jZ \cdot \tan \beta x \qquad (2.10)$$

where, $\beta = \omega\sqrt{LC}$ = the phase constant for the *transmission line*

          x = the length of the path from the box to the short.

Where βx is less than $\pi/2$ radians, i.e., when the electrical path

Sec. 2.3   Frequency Effects

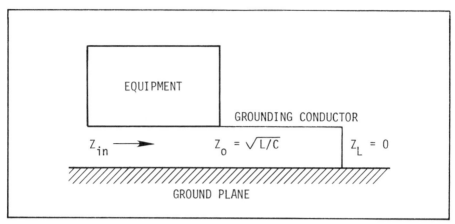

Figure 2.11 - Idealized Equipment Grounding.

length is less than $\lambda/4$, the input impedance of the short-circuited line is inductive with a value ranging from $0(\beta x=0)$ to ∞ ($\beta x=\pi/2$ radians). As βx increases beyond $\pi/2$ radians in value, the impedance of the grounding path cycles alternately between its open and short circuit values.

Thus, from the vantage point of the device or component which is grounded, the impedance is analogous to that offered by a short-circuited transmission line. Where $\beta x = \pi/2$, the impedance offered by the ground conductor behaves like a lossless parallel LC resonant circuit. Just below resonance, the impedance is inductive; just above resonance, it is capacitive; while at resonance, the impedance is real and quite high (infinite in the perfectly lossless case). Resonance occurs at values of x equal to integral multiples of quarter wave lengths, such as a half-wave length, three-quarter wave length, etc.

Typical ground networks are complex circuits of R's, L's and C's with frequency-dependent properties including both parallel and series resonances. These resonances are important to the performance of a ground network. Resonance effects in a groudning path are illustrated in Fig. 2.12. The relative effectiveness of a grounding conductor as a function of frequency is directly related to its impedance behavior (Fig. 2.13). It is evident from Fig. 2.12 that, for maximum efficiency, *ground conductor lengths should be a small portion of the wavelength at the frequency of the signal of concern*. Most effective performance is obtained at frequencies well below the first resonance.

Sec. 2.3 Frequency Effects

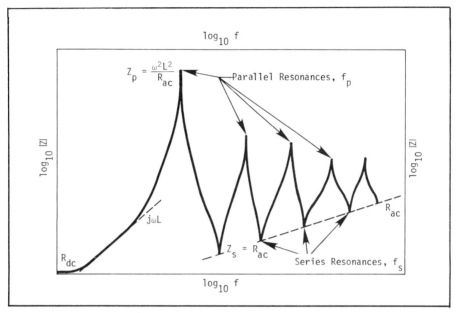

Figure 2.12 - Typical Impedance vs Frequency Behavior of a Grounding Conductor.

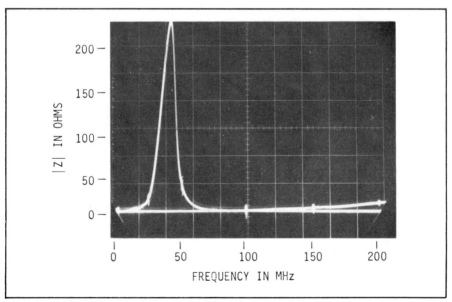

Figure 2.13 - Photograph of the Swept Frequency Behavior of a Grounding Strap.

Sec. 2.3   Frequency Effects

Antenna effects are related to circuit resonance behavior. Ground conductors will act as antennas to radiate or pickup potential interference energy, depending upon their lengths relative to a wavelength, i.e., their efficiency. This fact permits a wavelength-to-physical length ratio to be derived for ground conductors. The efficiency of a conductor as an antenna is related to its radiation resistance. Radiation resistance is a direct measure of the energy radiated from the antenna. A good measure of performance for a wire is a quarterwave monopole, which has a radiation resistance of 36.5 ohms.[6] An antenna which transmits or receives 10 percent or less than a monopole can logically be defined as inefficient. To be effective, a ground wire should be an inefficient antenna which means that it should exhibit a radiation resistance of 3.65 ohms or less. A monopole antenna with a length of less than $\lambda/11$ will have a radiation resistance of less than 3.65 ohms. A convenient criterion for a *poor* antenna, i.e., a good ground wire, is that its length be $\lambda/10$ or less. Thus, a recommended goal in the design of an effective grounding system is to maintain ground wires exposed to potentially interfering signals at lengths less than 1/10 of the interfering signal.

2.4 Common Impedance Coupling

Coupling can be defined as the means by which a voltage or current in one circuit induces a voltage or current in another circuit. Interference coupling is the stray or unintentional coupling between circuits which produces an unwanted response in one of the circuits. Since practical signal-reference planes do not exhibit a zero impedance, any currents flowing in such a plane will produce potential differences between various points on the reference plane. Interfacing circuits (equipment) referenced to these various points can experience conductively coupled interference in the manner illustrated in Fig. 2.14. The signal current, I_1, flowing in Circuit 1 returns to its source through signal reference impedance Z_R producing a voltage V_{N1} in the reference plane. The impedance Z_R is common to Circuit 2. Hence, V_{N1} appears in Circuit 2 as a voltage in series with the desired signal voltage source, V_{S2}. Similarly, the desired current, I_2, in Circuit 2 may produce interference in Circuit 1.

The conductive coupling of interference through the signal reference plane of interfaced equipment can occur in a manner similar to that described above for internal circuitry. If Circuit 1 represents two pieces of paired equipment and if Circuit 2 represents a

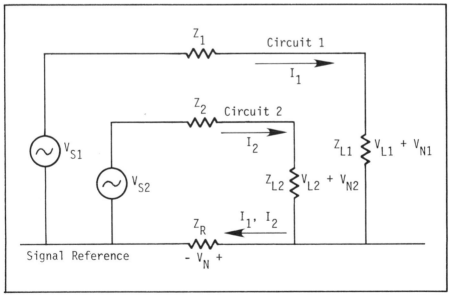

Figure 2.14 - Coupling Between Circuits Caused by Common Return Path Impedance.

Sec. 2.4 Common Impedance Coupling

different pair of interfaced equipment, then a current flowing in either circuit may produce interference in the other circuit as described.

Even if the equipment pairs do not use the signal reference plane as the signal return, the signal reference plane can still be the cause of coupling between them. Figure 2.15 illustrates the effect of a stray current, I_R, flowing in the reference plane. The current I_R may be the result of the direct coupling of another equipment pair to the signal reference plane. It may be the result of external coupling to the signal reference plane, or induced in the ground plane by an incident field. In either case, I_R produces a voltage V_N in the reference plane impedance, Z_R. This voltage produces a current in the interconnecting loop which in turn develops a voltage across Z_L in Equipment B. Thus, it is evident that interference can conductively couple through the signal reference plane to all circuits and equipment connected across the non-zero impedance elements of that reference plane.

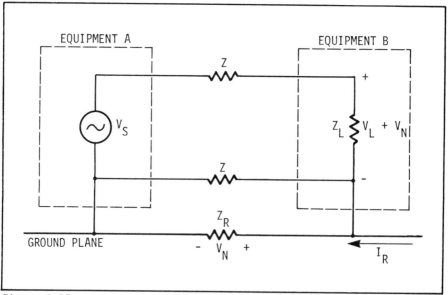

Figure 2.15 - Conductive Coupling of Extraneous Noise into Equipment Interconnecting Cables.

2.5 Interference Coupling

The basic mechanisms of interference coupling are summarized in Fig. 2.16, where (a) shows the signal return path of the circuit as sharing the ground, or reference plane, with other circuits or systems. These other currents will produce some voltage drop providing a threat of interference. Similarly, in circuit (b), the source and load ends of the signal transfer loop are shown connected to points of different potentials. Interference-threatening currents will thus be induced in the loop by the difference in potential. Situations (a) and (b) primarily are two views of the same problem and were discussed in the preceding section. Situations (c) and (d) for capacitive and inductive coupling also exist. Illustrations (e) and (f) show two views of the antenna effect that can be experienced in grounding systems. They are included to help visualize the effects of attempted corrective measures. In (e), the signal transfer path is configured as a loop antenna to illustrate magnetic field pickup. In (f), one end of the signal transfer path is shown connected to ground. The length of the signal return path is, however, electrically long enough to allow an appreciable voltage to be developed by an incident EM wave between the load end of the path and ground. This voltage will cause a current to flow through the signal return conductor and possibly cause interference.

In Fig. 2.16, the various coupling mechanisms, or modes, are illustrated separately. However, the usual interference control problem must deal with multiple effects simultaneously to assure that while one coupling effect is being suppressed another is not being enhanced. The relative importance of one mode versus another is a function of the desired signal circuit, system properties, and the nature of the spectral properties of the interference signal. The variation of this inter-relationship with system design and the environment makes it very difficult to formulate definitive rules of grounding which would be applicable to all circumstances.

Beyond reducing the magnitude of the noise-producing sources, or alteration of the desired signal's form, interference control is primarily a matter of reducing the coupling between the source of the interference and the desired signal circuit. The techniques for the reduction of coupling include:

- minimizing the impedance of the reference, or ground, plane (appropriate for Modes (a) and (b) of Fig. 2.16), if indeed the ground plane must be used for the return path.

- increasing the separation between the coupled circuits Modes (c) and (d)

- shielding the susceptible circuit or the source

Sec. 2.5 Interference Coupling

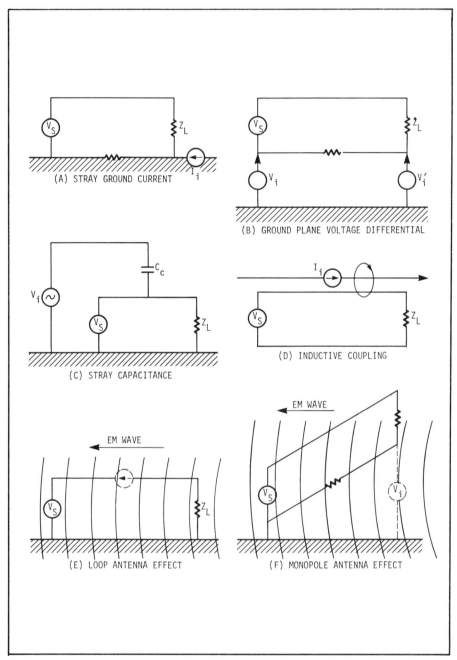

Figure 2.16 - Summary of Coupling Modes.

Sec. 2.5 Interference Coupling

Modes (c) through (f)

- reducing the area of signal loop Modes (d) and (e)

- balancing the signal transfer path (all Modes)

- electrically breaking the signal transfer loop at the frequency of the interference (all Modes)

- preventing the source and load ends of the desired signal circuit from being connected to points of different potential Modes (a) and (b)

- minimizing the efficiency of ground leads acting as antennas Modes (e) and (f)

- combinations of the above

These means for the reduction of coupling generally fall into the categories of:

(1) increased physical separation
(2) improved shielding
(3) ground-plane impedance minimization
(4) *breaking the ground loop* by some means.

Of these four, category (1) is self-evident and (2) is more appropriately treated elsewhere. Categories (3) and (4), however, are integrally related to ground path design and performance and are discussed in detail in Chapter 3.

2.6 References

1. Denny, H.W., et al, *Grounding, Bonding and Shielding Practices and Procedures for Electronic Equipments and Facilities*, Vol. I, Fundamental Considerations, Engineering Experiment Station, Georgia Institute of Technology, Atlanta, Ga., Report No. FAA-RD-75-215, I, AD-A022-332, December 1975.

2. Terman, F.E., *Radio Engineer's Handbook*, McGraw-Hill Book Company, Inc.; New York, NY (1943).

3. Troup, R.J., and Grubbs, W.C., *A Special Research Paper on Electrical Properties of a Flat Thin Conductive Strap for Electrical Bonding*, Proceedings of the Tenth Tri-Service Conference on Electromagnetic Compatibility, IITRI, Chicago, Il., November 1964, pp. 450-474.

4. White, D.R.J., *EMI Control Methodology and Procedures*, Don White Consultants, Inc., Gainesville, Va., (1981).

5. Kraus, J.D., *Electromagnetics*, McGraw-Hill Company, Inc., New York, NY (1953).

6. Jordan, E.C., *Electromagnetic Waves and Radiating Systems*, Prentice-Hall, Inc., Englewood Cliffs, NJ (1950).

CHAPTER 3

Control of Unwanted Coupling

In the previous chapter, interference between circuits and systems was related to the coupling which always exists between ground conductors. Grounding conductors and paths exhibit inductive, capacitive, and resistive properties. Having such properties means that grounding conductors and paths rarely provide the zero impedance, or equipotential, reference plane which is sought for the grounding of signals. Any unwanted signals, whether they be power or microwaves, pose a threat of interference to desired signals whose circuits are referenced to the network. In this chapter, techniques for minimizing unwanted coupling to and from signal circuits are described.

3.1 Ground Plane Impedance Control

If both source and load ends of a circuit are connected to a noisy ground reference, a possibility exists for interference. Obviously, if the voltage differential between the source and the load is reduced, the interference threat is reduced. Frequently, reduction of interference can be done through lowering the impedance of the path through which the interference currents (and, frequently, the desired signal currents) are flowing. Ground noise problems can frequently be solved by providing more ground paths through the bonding of all metallic members of an equipment cabinet, or of a structure, together with low impedance interconnections. For example, Fig. 3.1 illustrates the application of this principle in the bonding of cable trays and Fig. 3.2 shows it applied to structural members. The technique, if thoroughly applied, throughout a facility, can be **very** effective in reducing ground system (plane) impedance. For example, in one instance[1] the overall background noise in a space system checkout facility was reduced enough to allow the rescheduling of EMI and system tests from nightime to regular daytime hours. The use of massive ground-return conductors, grounding sheets, or large area plates are frequently helpful. Figures 3.3 and 3.4 are illustrations of the use of large grounding conductors to reduce the voltage differences arising from potentially interfering signals

Sec. 3.1	Ground Plane Impedance Control

Figure 3.1 - Multiple Grounding of Cable Trays.

Figure 3.2 - Structural Bonding.

Sec. 3.1   Ground Plane Impedance Control

Figure 3.3 - Example of Large Ground Busses.

Figure 3.4 - Application of Large Ground Busses to Control Interference.

Sec. 3.1   Ground Plane Impedance Control

sharing a common ground.

Radio Frequency (RF) equipment and circuits frequently utilize the chassis, cabinet, or large unetched areas on printed circuit boards as ground planes for the signals of primary interest. Large, common areas are effective RF signal grounds because wide metal paths exhibit lower inductance, and thus lower impedance, than do round wires or narrow rods. To retain the ground plane effectiveness, paths serving as individual component, device, or network grounds must be short to minimize the circuit-relative impedance; thus, the overall ground must be *brought to* the component, device, or network. Radio frequency interfaces (input and output signal ports) are typically unbalanced and thus cable references, or signal grounds, must be the same as the chassis or cabinet ground.

Attempts to realize an adequately low ground plane impedance, i.e., minimize ground reference voltage differentials, is the basis of multiple-point grounding. In multiple-point grounding, the source and load end of every energy (signal) transfer pair is connected to a ground plane (chassis, equipment cabinet, structural frame, printed circuit board, etc.) by the shortest electrical path. The ground plane is also interconnected by the largest feasible number of parallel paths. Ideally, the ground plane would consist of a solid metal mass of lowest possible resistance.

3.2 Opening Ground Loops[2]

Whenever interference coupling involves the ground plane or return, a *ground loop* is said to exist. To counter the effects of a ground loop, it is necessary to provide some form of discrimination against the ground-path related interference. This discrimination, or rejection, is most commonly provided through:

- single-point grounding
- common-mode rejection
- frequency translation
- optical coupling
- frequency-selective grounding

3.2.1 Single-Point Grounding

Single-point grounding means connecting one point (either the source or load end) of the signal return side of the energy transfer loop to ground. There are many applications which require single-point grounding. Although definitely advantageous in many situations, in a multiequipment system or facility a single-point ground systems can be very difficult to implement and maintain as a true single-point ground system. The pros and cons of single-point grounding are discussed in more detail in the next chapter.

3.2.2 Common-Mode Rejection

When both ends of a signal loop must be grounded to a noisy ground plane, the resultant differential voltage can cause the flow of common-mode currents. Balanced circuits are an effective way to discriminate against such common-mode voltages and currents. Complete rejection of common-mode interference requires perfect balance between both legs of the circuit. Generally, a high degree of balance becomes more difficult with increasing frequency.

Figure 3.5 shows a typical balanced (differential) line. Such circuits can generally achieve up to 70 dB of common-mode rejection at frequencies up to several hundred kHz.

Figure 3.6 shows balanced operation achieved with balanced-to-unbalanced transformers. Effective operation over broad frequency ranges can be achieved with adequate transformer cores and proper windings. A limitation of the circuit in Fig. 3.6 is that it will not transmit dc. The use of common-mode chokes as in Fig. 3.7 will transmit dc and differential mode ac signals while rejecting common-mode ac signals.

Sec. 3.2   Opening Ground Loops

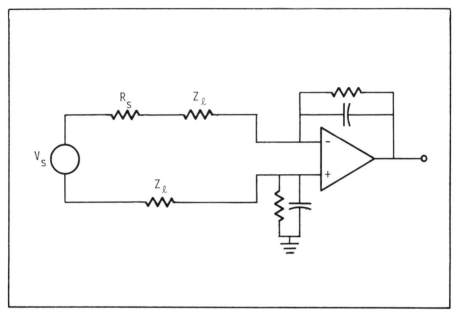

Figure 3.5 - Use of Differential Line Receivers for Common-Mode Rejection.

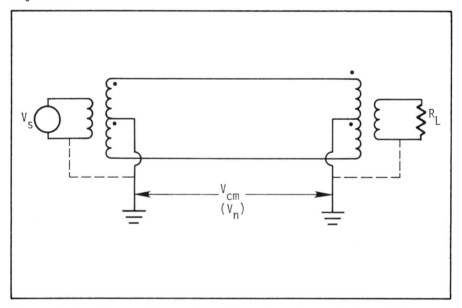

Figure 3.6 - Balanced Operation with Balanced-to-Unbalanced Transformers.

Sec. 3.2   Opening Ground Loops

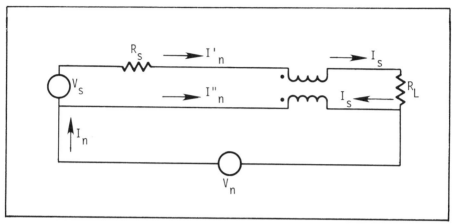

Figure 3.7 - Common-Mode Chokes.

3.2.3 Frequency Translation

When the common-mode signals are essentially at the same frequency as those of the desired signals, effective discrimination can be obtained by frequency translation of the desired signal. This involves altering the form of the desired signal by chopping, modulating the desired signal on a carrier, or other means.

3.2.4 Optical Isolation

Isolation via optical coupling (i.e., optical isolators or fiber optics) is a very effective means of achieving common-mode rejection. It is particularly useful where potentially hazardous voltages may exist between ground points. Optical coupling cannot be used for low-level analog circuits.

Figure 3.8 shows the use of optical isolation to provide common-mode rejection. Note that the noise voltage V_n does not enter the desired signal path. Also, note that previously described steps must be used to keep V_n from entering either of the source or load-related loops. Optical isolation is a special case of frequency translation in that the desired signal information is translated up to optical frequencies prior to transmission.

Sec. 3.2　Opening Ground Loops

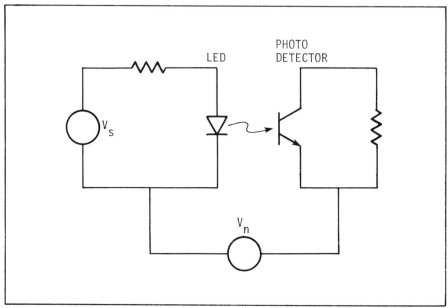

Figure 3.8 - Use of Optical Isolation to Combat Common-Mode Noise.

3.2.5　Frequency-Selective Grounding

Figures 3.9 and 3.10 are examples of specialized grounding techniques. In situations where single-point grounding is required at low frequencies while multiple point grounding is required at high (RF) frequencies, capacitive grounding can be used as shown in Fig. 3.9. The inverse situation is shown in Fig. 3.10 where an inductance is used to achieve a dc and low frequency ground (e.g., for safety) and approximate an open circuit at RF. Parasitic effects can render capacitive and inductive grounding highly unpredictable and should be used only under unusual conditions and with a great deal of care to assure that the overall interference problem is not aggravated.

Sec. 3.2   Opening Ground Loops

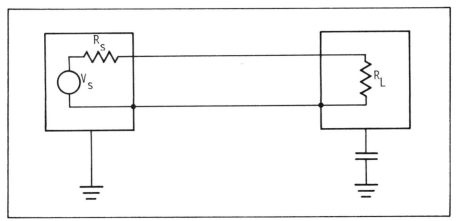

Figure 3.9 - Capacitive Grounding.

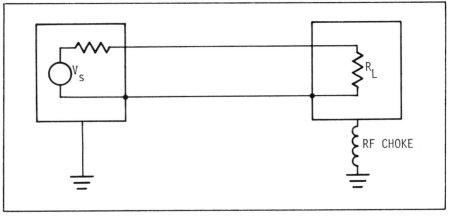

Figure 3.10 - Inductive Grounding.

3.3 References

1. Lightner, D.R. and Toler, J.C., *Implementation of Bonding Practices in Existing Structures,* Proceedings of the Eighth Tri-Service Conference on Electromagnetic Compatibility, Chicago, Il., October 1962, pp. 670-690.

2. White, D.R.J., *EMI Control Methodology and Procedures,* Don White Consultants, Inc., Gainesville, Virginia, (1981).

CHAPTER 4

Ground Network Configurations

If a truly zero impedance ground reference plane or bus could be realized, it could be utilized as the return path for all currents --power, control, audio and RF--present within a system or complex. Such a ground reference would provide the necessary fault protection simultaneously with lightning, static discharge, and signal return paths. The closest approximation to an ideal ground would be a very large sheet of conductor material, such as copper or aluminum, underlying the entire facility with large risers extending up to individual pieces of equipment--a prohibitively expensive approach. The *typical* approach is to utilize a network of wires, tubes, pipes, and bars of copper, aluminum or steel configured in a manner to produce reference *plane* characteristics most appropriate for the frequencies employed within the system.

Within a piece of equipment, signal ground network consist of a sheet of metal which serves as a signal reference plane for some or all of the circuits in that equipment. Between units of equipment, where units are distributed throughout the facility, the signal ground network usually consists of a number of inter-connected wires. Whether serving a collection of circuits within an equipment or serving several pieces of equipment within a facility, the signal ground network will be a floating ground, a single-point ground, or a multiple-point ground. This chapter compares features and applications of these three configurations.

4.1 Floating Ground

A floating ground is illustrated in Fig. 4.1. This type of signal ground system in a facility is electrically isolated from the building ground and other conductive objects. Hence, noise currents present in the ground system will not be conductively coupled to the signal circuits. The floating ground system concept is also employed in equipment design to isolate signal returns from equipment cabinets and thus prevent unwanted currents in cabinets from coupling directly to signal circuits.

Effectiveness of floating ground systems depends upon their true isolation from other nearby conductors, i.e., to be effective, floating ground systems must really float. In large facilities, it is often difficult to achieve a completely floating system, and even if complete isolation were achieved, it would be difficult to maintain such a system.[1] Such a floating system can be practical only if a few circuits or a few pieces of equipment are involved and power is supplied from either batteries or dc-to-dc converters.

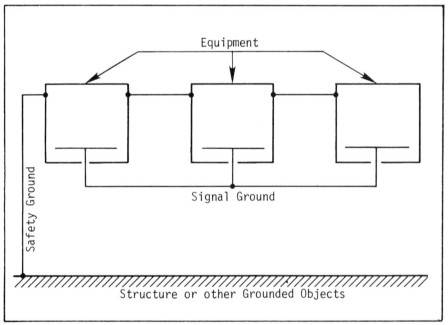

Figure 4.1 - Floating Signal Ground.

4.2 Single-Point Ground

In many situations, ground plane noise voltages pose a very definite interference threat to circuits and equipment referenced to the ground plane. In particular, those systems which operate at very low current or voltage levels and involve the processing of low frequency analog information are generally susceptible to ground network noise. Examples of particularly susceptible systems include nuclear reactor control monitors, industrial process controllers, baseband video systems, audio communications, and physical parameter sensors such as thermocouples, strain gauges, accelerometers, etc.

Much of the noise present in a structural ground system of a facility (building, ship, vehicle, etc.) is usually traceable to the primary power (50, 60, or 400 Hz) system. Primary power enters the ground network directly as a result of improper wiring (or improper design, as in the case of a severely unbalanced three phase wye - wye system) or indirectly from capacitor leakage (or coupling), insulation leaking, or magnetic induction. Where the susceptible bandwidth of the desired signal processing network encompasses either the primary frequency of the power signal, or some of its harmonics, the threat of interference is high. In considering the potential susceptiblity of a circuit or network to ground system noise, noise components produced by relays, solenoids, rectifiers, etc., should also be considered.

Another situation which seems to call for single-point grounding is where protection against EMP* fields is required.

A single-point ground for an equipment complex is illustrated in Fig. 4.2. With this configuration, the signal circuits are referenced to a single point, and this single point is then connected to the facility ground. The ideal single-point signal ground network is one in which separate ground conductors extend from one point on the facility ground to the return side of each of the numerous circuits located througout a facility. This type of ground network requires an extremely large number of conductors and is not generally economically feasible. In lieu of the ideal, various degrees of approximation to single-point grounding are employed.

The configuration illustrated by Fig. 4.3 represents one approach to a facility-wide single-point ground. It uses individual ground buses extending from a single point on the facility ground to

---

* Electromagnetic Pulse-EMP is a very short burst of highly intense electromagnetic energy produced by nuclear detonations. It is characterized by a very short (10 nsec) risetime, high peak field intensity (50 kV/m), and a short duration (250 nsec). Most of the energy is concentrated at frequencies below 10 MHz, where magnetic field effects predominate.

Sec. 4.2   Single Point Ground

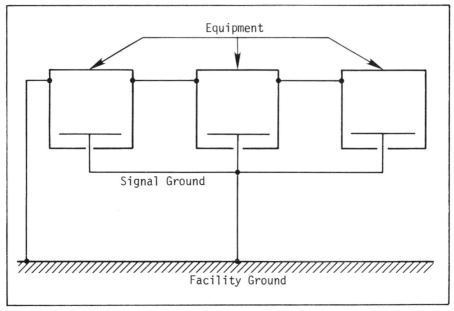

4.2 - Single-Point Signal Ground.

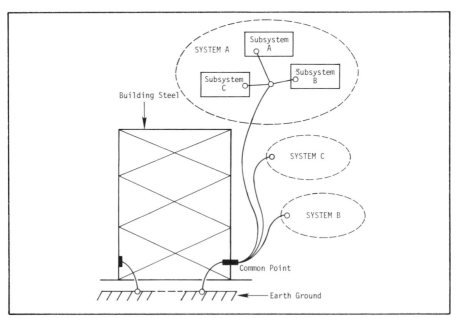

Figure 4.3 - Single-Point Ground Bus System Using Separate Risers.

Sec. 4.2 Single Point Ground

each separate electronic system. In each system, the various electronic subsystems are individually connected at only one point to the ground bus. The ground bus network arrangement illustrated in Fig. 4.4 assumes the form of a tree. Within each system, each subsystem is single-point grounded. Each of the system ground points is then connected to the tree ground bus with a single insulated conductor.

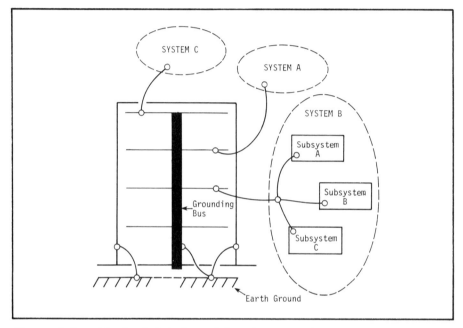

Figure 4.4 - Single-Point Ground Bus System Using a Common Bus.

Electromagnetic pulse protection methodology[2] calls for a single-point ground configured as illustrated in Fig. 4.5. In this zonal[3] concept of grounding, one and only one ground connection interconnects the zonal boundaries. No ground connection passes through a zonal boundary; it must terminate at the boundary and internal grounds to a zone begin at the boundary. Zone 0 usually means the region outside of the facility (out-of-doors). Zone 1 is the mildly attenuated region just inside facility walls. Zone 2 is the further attenuated region inside a room or an enclosure. Zones 3 and higher exist inside equipment enclosures or in compartments within equipment. A zonal boundary is any well-defined and controlled interface or wall that offers some degree of electromagnetic shielding to an electromagnetic wave in zone 0.

4.5

Sec. 4.2   Single Point Ground

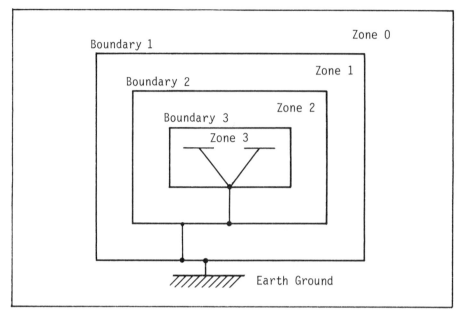

Figure 4.5 - Zonal Grounding.

The single-point ground accomplishes each of the three functions of signal circuit grounding. That is, a signal reference plane is established in each unit or piece of equipment and these individual reference planes are connected together. These, in turn, are connected to the facility ground at least at one point which provides fault protection for the circuits and provides control over static charge buildup.

An important advantage of the single-point configuration is that it helps control conductively-coupled interference. As illustrated in Fig. 4.6, closed paths for noise currents in the signal ground network are avoided, and the interference voltage, V_N, in the facility ground system is not conductively coupled into the signal circuits via the signal ground network. Therefore, the single-point signal ground network minimizes the effects of any noise currents which may be flowing in the facility ground.

In a large installation, a major disadvantage of a single-point ground configuration is the requirement for long conductors. In addition to being expensive, long conductors prevent realization of a satisfactory reference plane for higher frequencies because of large self-impedances. Further, because of stray capacitance between conductors, single-point grounding essentially ceases to exist as the signal frequency is increased.

Sec. 4.2   Single Point Ground

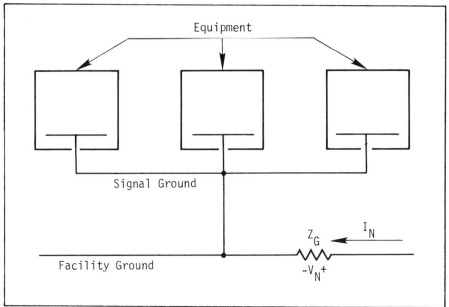

Figure 4.6 - Use of Single-Point Ground Configuration to Minimize Effect of Facility Ground Currents.

Strict attention must be paid through all phases of design, installation, operation, and maintenance to assure that the single-point configuration is retained to realize the full advantages offered by single-point grounding. Interfacing circuits (paths) between elements or equipments grounded to separate branches of the tree must employ one of the techniques described in Chapter 3 to assure that the separate ground points remain isolated except through the one path.

4.3 Multiple-Point Ground

The multiple-point ground illustrated in Fig. 4.7 is the third configuration frequently used for signal ground networks. This configuration establishes many conductive paths to various electronic systems or subsystems within a facility. Within each subsystem, circuits and networks have multiple connections to this ground network. Thus, in a facility, numerous parallel paths exist between any two points in the ground network as shown in Fig. 4.8.

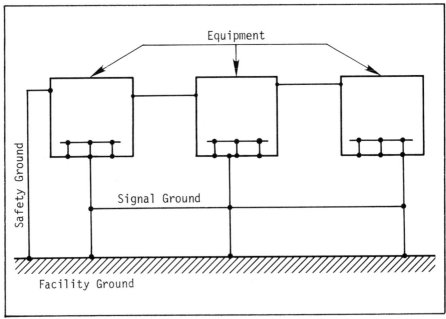

Figure 4.7 - Multiple-Point Ground Configuration.

Multiple-point grounding frequently simplifies circuit construction inside complex equipment. It permits equipment employing coaxial cables to be interfaced more easily since the outer conductor of the coaxial cable does not have to be floated relative to the equipment cabinet or enclosure.

However, multiple-point grounding suffers from an important disadvantage. Sixty hertz power currents and other high amplitude low frequency currents flowing through the facility ground system can conductively couple into signal circuits to create intolerable interference in susceptible low-frequency circuits. Also, ground loops

Sec. 4.3   Multiple-Point Ground

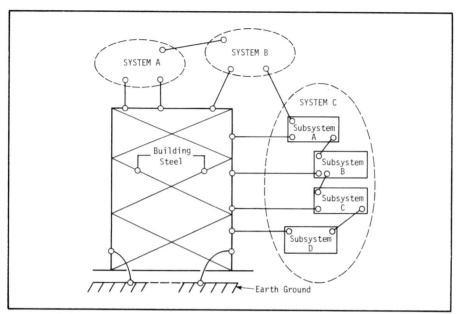

Figure 4.8 - Use of Structural Steel in Multiple-Point Grounding.

are created in many places which makes them more prone to radiated pickup.

4.3.1 Requirements

For multiple-point grounding to be effective, the following conditions must exist:

- All grounding conductors between the separate points desired at the same reference *must be less than 0.1 wavelength* of the interference or noise-producing signal. Otherwise, inductance, resonance effects, and *antenna pickup* will prevent the zero-potential condition from being realized. Where relatively short distances are involved, as on a circuit board for instance, path inductance and stray capacitance usually can be controlled sufficiently to allow multiple-point grounding to be effectively achieved. Even so, care in design is in order because lead inductances can resonate with various stray capacitances and produce unexpectedly high and low impedance paths. Such paths may produce unintentional ringing, offer unanticipated sneak paths, cause excessive feedback, etc.

Sec. 4.3 Multiple Point Ground

- A second condition which must be met for multiple-point grounding to be effective is that the voltage differentials on the ground plane must be reducible by the added ground paths. For example, the voltage differentials existing within a steel frame building from stray power currents are not likely to be significantly affected by supplying additional grounding conductors between the equipment elements distributed throughout the building. To exert a noticeable influence, the combined impedance of the added ground conductors should be at least an order of magnitude lower than that already existing between all points of the existing ground system. The criterion is actually based on that required to insure that any interference is below victim sensitivity. Frequently, the realization of such a low impedance is not feasible or economical. As an example, assume that two pieces of equipment (e.g., data processors), although subsets of a larger system, are located on different floors of a building as shown in Fig. 4.9. Further assume that the steel frame building is effectively bonded at the joints so that a low *resistance* path exists throughout. If independently supplied with power, each equipment cabinet will have a safety ground that will likely put each cabinet in contact with the structure near their respective locations. Any ΔV from stray power currents, radiant RF sources, etc., existing between these two structural locations are not likely to be materially affected by any reasonably-sized auxilliary ground bus between the two pieces of equipment. The reason is that structural members, although of high resistivity materials, typically offer enough cross-sectional area to yield a lower impedance than any practical grounding bus.[4]

Sec. 4.3   Multiple-Point Ground

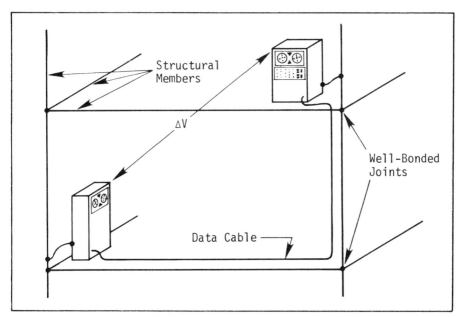

Figure 4.9 - Multiple Equipment Grounding.

4.3.2 Applications

Multiple-point grounding is commonly employed under the following general situations:

- *Circuit construction.* Where many components and devices (resistors, capacitors, transistors, diodes, transformers, etc.) are interconnected in a complex manner, it is not possible to segregate every source load pair. Also, it is not practical to supply power to each with separate conductors. Circuit board grounds, which frequently also serve as dc returns represent a common example of a multi-point ground as shown by Fig. 4.10.

- *RF Equipment.* As noted earlier, equipment utilizing RF* signals typically employ unbalanced (television receivers being the notable exception) transmission lines to bring signals in and to carry them out.

---

* From the perspective of grounding, the frequency at which RF begins is subject to considerable controversy. This question will be addressed later.

Sec. 4.3   Multiple-Point Ground

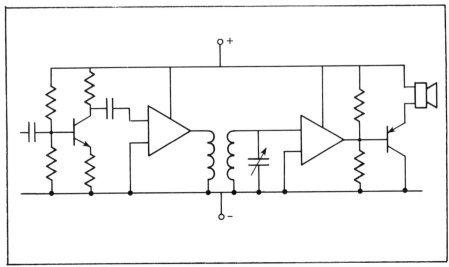

Figure 4.10 - Common Multiple-Point Ground.

- To maintain effective shield performance and a low impedance return for the transmission line signal, one side of the signal interface (coaxial connector, waveguide flange, etc.) is at *ground*, i.e., chassis or cabinet, potential. Normal construction dictates that the signal return side of such interfaces be integrally bonded to the equipment chassis. Otherwise, unusual procedures in construction and painstaking care to prevent future compromises may be required.

- *High Level RF Exposure.* In high level RF environments, such as near a broadcast transmitter, it is desirable to minimize the RF potentials existing between interconnected equipments. The use of extensive ground planes with multiple bonds between this ground plane and equipment cabinets, and between the equipment enclosures themselves, is frequently needed to control excess pickup of the unwanted RF. Another reason for multiple interconnections is to eliminate RF-induced arcs which can be both hazardous and a source of broadband RF noise.

- *Power Safety.* Exposed metal surfaces enclosing

energized conductors above certain voltages must be grounded for personnel safety. This power safety ground usually involves structural metals, utility pipes, raceways, conduit, etc. This interconnected complex generally evolves into a multiple-point ground. (A multiple-point ground, however, is not necessary for personnel safety, per se.).

- *Lightning Protection.* The fast risetime, high amplitude current pulse created by a lightning discharge typically generates very high voltages between conductors. Such voltages may easily exceed the arc-over potential of air or other intervening dielectric material and produce flashover. Such occurrences are most easily prevented by crossbonding all conducting objects within flashover distances of the main lightning conductors.

4.4 Recommended Approach

Because of the interference threat that stray power currents pose to audio, digital, and control circuits (or others whose operating bandwidths extend down to 60 Hz or below), steps must be taken to isolate such large currents from signal return paths. Obviously, one method of lessening the effects of large power currents is to configure the signal-ground system so the signal return path does not share a path in common with a power return.

The first step in the development of an interference-free signal reference system for equipment, or a facility, is to assure that the ac primary power return lines are interconnected with the safety grounding network only at one point. Isolation of ac power returns from the signal reference network is very effective in reducing many noise problems. However, additional steps should be taken to minimize other stray ac currents such as those resulting from power-line filters. One method of reducing such currents is to limit the number of filter capacitors in an installation by using commonly filtered ac lines wherever possible.

To meet the safety requirements, while minimizing the effects of power currents flowing with signal currents through a common impedance, a single connection* between the signal reference ground network and building ground is necessary. This single connection eliminates conductive loops in which circulating (low frequency) currents can flow to produce interference between elements of the network. However, the inductance of the long conductors frequently associated with a single-point ground prevents the realization of a satisfactory reference plane at high frequencies.

Unfortunately, because of the large number of closed loops in a multiple-point ground, power frequency currents and other low frequency sources present a serious noise threat to low frequency equipment. Therefore, multiple-point grounding should not be used for low frequency systems.

Thus, neither single-point grounding nor multiple-point grounding alone is able to provide the reference plane characteristics desired over a wide range of frequencies. Therefore, a hybrid

---

* The connection can be made at any location within a facility. Tradition favors making the connection at the point where the lightning protection system and the power fault system attach to the earth electrode system. However, there seems to be no solid evidence that this point offers any significant advantage over another insofar as the elimination of interference is concerned, but some advantage is offered against lightning flashover by connecting the signal ground directly to the earth electrode system.

Sec. 4.4   Recommended Approach

approach is usually necessary that implements the single-point principle at low frequencies and the multiple-point principle at high frequences.

4.4.1 Low-Frequency Network

A low-frequency grounding network for a facility should conform to the following principles:

- It should be isolated from other ground networks including structural grounds, safety grounds, electrical equipment grounds, etc.

- An inter-equipment or facility grounding network should not be used to provide the primary return path for signal currents from the load to the source.

- A low-frequency grounding network must be connected to the facility ground (and thence to the earth) at only one point. This connection can be made at the point where the lightning protection system and the power fault protection system attach to the earth electrode system. Where this point is not accessible, select a convenient point.

- A network must be configured to minimize conductor path lengths. In facilities where the equipment elements (or chassis) to be connected to the ground network are widely separated, more than one network should be installed.

- Finally, conductors of the network should be routed in a manner that avoids long runs parallel to primary power conductors, lightning down conductors, or any other conductor likely to be carrying high amplitude currents.

4.4.2 High-Frequency Network

In high-frequency systems, equipment chassis frequently are used as the signal reference. The chassis in turn usually is connected to the equipment case at a large number of points to achieve a low impedance path at the frequencies of interest.

The National Electrical Code requires that equipment cases and housings be grounded to protect personnel from hazardous voltages in the event of an electrical fault. Stray currents in the fault protection network can present an interference threat to any signal

Sec. 4.4   Recommended Approach

system whose operating range extends down into the power and low-frequency range. Where such problems exist, it is advisable to attempt to reduce the impedance of the reference plane as much as possible. One approach is to interconnect equipment enclosures with building structural steel, cable trays, conduit, heating ducts, piping, etc., into a facility ground system to form as many parallel paths as possible.* It should be recognized that because of the inductance and capacitance of network conductors, such multiple-point ground systems offer a low impedance only to the low frequency noise currents; however, these currents can be most troublesome in many facilities.

4.4.3 Selection of the Dividing Frequency

If one type of ground system is appropriate for low frequency signals and another is appropriate for high frequency signals, a dividing line between high and low frequencies must be defined. In addition, the signals of primary concern must be established. The signals of concern may be environmentally related or may be associated with normal equipment/system functioning and related to basic data rates, baseband signal ranges, carrier frequencies, etc.

In facilities, the actual operating signals within a system are more definable than the external environmental threat. A general idea of the probable types of equipment that are to be installed in a facility will normally be available. It is known in advance that certain types and classes of equipment (such as telephone circuits) are more likely to be susceptible to power frequency currents and voltages than other types (such as RF systems). Relative to the external environment, RF sources, such as radio transmitters and radars, are capable of causing interference. However, the specific RF environment is typically not known prior to facility construction.** Therefore, tailoring the dividing frequency to the external environment at different locations is not generally realistic. Consequently, the dividing line between high and low frequency is usually based upon the recognized problem posed by low-frequency interference sources (primarily stray power currents and fields) to low frequency systems.

Table 4.1 lists several frequencies that have been used or suggested as the dividing line between low and high. Practically all

---
* Since many ground loops are so formed, multiple-signal return paths can be avoided by floating motherboards from their chassis and/or using shields of many skin depths for conductive isolation.
** It is strongly recommended that an RF survey be conducted *prior to selecting the site* so that appropriate design measures can be implemented to enhance the compatibility of the final system with the environment.

4.16

Sec. 4.4   Recommended Approach

Table 4.1 - Various Suggested Dividing Frequencies Between
Single Point and Multiple Point Grounding

Frequency in MHz	Source Reference
0.03/0.3	5
0.05	6, 7
0.1	8, 9
0.15	10
1	11, 12
2	4

have tended to focus on the primary operating frequency range of a system with the idea that, within these ranges, the associated circuits are most sensitive and thus most vulnerable to extraneous signals of comparable frequencies.

4.5 References

1. Denny, H.W., et al, *Electronic Facility Bonding, Grounding and Shielding Review*, Report No. FAA-RD-73-51, Department of Transportation, Washington, D.C. 20591, November 1972.

2. DNA EMP (Electromagnetic Pulse) Handbook, DNA 2114-1, Defense Nuclear Agency, Washington, D.C., November 1971.

3. Vance, E.F., *Shielding and Grounding Topology for Interference Control*, EMP Interaction Note 306, SRI International, Menlo Park, Ca., April 1977.

4. *Final Report on the Development of Bonding and Grounding Criteria for John F. Kennedy Space Center*, WDL-TR 4202 (3 Volumes), Contract NSIO-6879, Philco-Ford Corp., Palo Alto, Ca., 30 June 1970.

5. MIL-STD-124-188, *Grounding, Bonding and Shielding for Common Long Haul/Tactical Communication Systems*, 14 June 1978.

6. *Electromagnetic Interference Specification for Apollo*, Contract No. NAS9-150, Genistron Incorporated, L.A., Ca., 15 February 1962.

7. *Performance Specification for Equipment Grounding Requirements on Preflight Acceptance Checkout Equipment - Spacecraft (PACE - s/c) Program*, NASA/MSC PACE-s/c Project Office Specification 53, Revision 1, Cape Canaveral, Fl., 15 November 1963.

8. Coge, J.R., *Electromagnetic Compatibility Requirements for Space Systems*, TOR-1001 (2307)-4, Reissue B, Contract F04701-72-C-0073, Aerospace Corp., El Sequndo, Ca., 15 August 1972.

9. *RFI Control Plan for Project Gemini Spacecraft*, Genistron Inc., L.A., Ca., 24 August 1962.

10. *Electrical-Electronic Grounding Plan-Saturn S-IC*, Document D5-11207, Contract NAS8-2577, Boeing Company, Seattle, Wa., 1 August 1963.

11. Taylor, R.E., *Radio Frequency Interference Handbook*, NASA-SP-3067, National Aeronautics and Space Administration, Washington, D.C., 1971, N72-1153-156.

12. Denny, H.W., et al, *Grounding, Bonding and Shielding Practices and Procedures for Electronic Equipments and Facilities*, (3 Volumes), Report No. FAA-RD-75-215, Contract No. DOT-FA72WA-2850, Engineering Experiment Station, Georgia Institute of Technology, Atlanta, Ga., Dec. 1975, AD A022 332, AD A022 60 8, and AD A022 871.

CHAPTER **5**

Grounding for Fault Protection

Fault protection is an integral part of ground network design. Frequently, signal grounds are common with fault protection paths, particularly where structural elements of a building are involved. Because traditional practices and personnel and structural protection requirements strongly favor that priority be placed on fault protection needs over EMI needs, it is likely that signal ground networks must be designed to be compatible with these networks. Occasionally, requirements of fault protection will force compromises in the signal ground design and implementation. The following discussion presents an overview of the principles of design behind fault protection systems.

5.1 Electric Shock

Electric shock occurs when the human body becomes a part of an electric circuit. It most commonly occurs when people come in contact with energized devices or circuits while touching a grounded object or while standing on a damp floor. The effects of an electric current on the body are principally determined by the magnitude of current and duration of the shock. Current is determined by the open circuit voltage of the source and total path resistance including internal source resistance and human body resistance. In power circuits, internal source resistance is usually negligible in comparison with that of the body. In such cases, the voltage level, V, is the important factor in determining if a shock hazard exists. At commercial frequencies of 50-60 Hz and voltages of 120-140 volts, the contact resistance of the body primarily determines the current through the body. This resistance may decrease by as much as a factor of 100 between a completely dry condition and a wet condition. For estimation purposes, the resistance of the skin is usually somewhere between 500 and 1500 ohms.

An electric current through the body can produce varying effects including death, depending upon the magnitude of current.[1]. For

Sec. 5.1   Electric Shock

example, the perception current is the smallest current that might cause an unexpected involuntary reaction and produce an accident as a secondary effect. Shock currents greater than the reaction current produce an increasingly severe muscular reaction. Above a certain level, the shock victim becomes unable to release the conductor. The maximum current at which a person can still release a conductor by using the muscles directly stimulated by that current is called the *let-go* current. Shock currents above the *let-go* level can begin to cause chest muscles to contract and breathing to stop. If the current is interrupted quickly enough breathing will resume. At a still higher level, electric shock currents can cause an effect on the heart called ventricular fibrillation. Under this condition, usually there is a stoppage of heart action. Various current levels for 60 Hz and for dc are summarized in Table 5.1. At frequencies above 300 H_z, the current levels required to produce the above effects begin to increase due to skin effect. For example, the perception current is approximately 100 mA at 70 kHz. Above 100-200 kHz, the sensation of shock changes from tingling to heat. It is believed that heat or burns are the only effects of shock above these frequencies.

Table 5.1 - Summary of the Effects of Shock[2]

Alternating Current (60 Hz)	Direct Currents	Effects
(mA)	(mA)	(mA)
0.5-1	0-4	Perception
1-3	4-15	Surprise (Reaction Current)
3-21	15-80	Reflex Action (*Let-go* Current)
21-40	80-160	Muscular Inhibition
40-100	160-300	Respiratory Block
Over 100	Over 300	Usually Fatal

5.2 Fault Protection Objectives

To protect people from inadvertant exposure to hazardous voltages, all exposed metallic elements should be connected to *ground*. In this sense *ground* usually means other exposed equipments, metal members of the building, plumbing fixtures, and any other metallic sturctures likely to be at a different (hazardous) potential in the event of a fault. Then, if accidental contact occurs between energized conductors and chassis, frame, or cabinet through human error, insulation failure or component failure, a direct low resistance path exists between the fault and the energy source (usually a transformer) which causes fuzes to blow or breakers to trip and thus quickly remove the hazard.

The previously noted dependence of shock effects upon frequency suggests that the relative danger from electric shock is related to the duration of exposure. Reference 2 indicates that the current levels required to produce a given effect are inversely proportional to the square root of the time of exposure. Thus, one objective of the fault protection system (network) is to reduce the time of exposure to a minimum. Another reason for achieving rapid clearance is to limit temperature rise in the faulted conductor and thus minimize a potential fire hazard.

5.3 Fault Protection Design

The National Electric Code (NEC) is issued by the National Fire Protection Association as NFPA 70 and is updated every three years. This code has been adopted by the American National Standards Institute as ANSI C1 and it is rapidly becoming universally accepted. It has become, in effect, a Federal Standard with its incorporation into the Occupational Safety and Health Administration (OSHA) requirements. When viewed from the perspective of EMI and noise control, it should be remembered that the primary objectives of the NEC are fire and shock hazard protection and not the achievement of electromagnetic compatibility. However, this does not mean that practices of the NEC are necessarily in opposition to good EMI practices. It does mean that the NEC standards must be met in any equipment or system design and installation and therefore its requirements must be thoroughly understood while implementing any EMI grounding system. Another thing to realize is that simply conforming to the NEC does not assure a noise-free system; generally other measures also must be employed to achieve the desired level of total system (equipment, facility, and environment) compatibility.

Article 250 of the NEC sets forth the general grounding requirements for electrical wiring in a structure. Present requirements of the NEC specify that the ground lead (green wire), in a single-phase ac power distribution system, must be one of three leads. The other

Sec. 5.3   Fault Protection Design

two leads comprise the *hot* lead (black wire) and the *neutral* lead (white wire). The ground lead is a safety conductor designed to carry current only in the event of a fault. The *hot* lead is connected to the high side of the secondary of the distribution transformer. For fault protection, the NEC specifies that the neutral be grounded at the service disconnecting means (main breaker). The safety ground (green wire) is grounded at that point as shown in Fig. 5.1. All exposed metallic elements of electrical and electronic equipment are connected to this ground with the green wire.

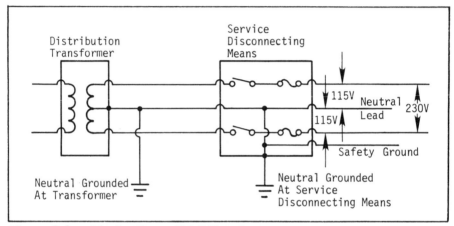

Figure 5.1 - Single Phase 115/230 Volt AC Power Ground Connections.

Grounding of a three phase wye power distribution system is done similarly to the single-phase system. The connections for a typical system are shown in Fig. 5.2. The neutral lead, as in single-phase systems, is grounded for fault protection at the service disconnecting means. The NEC specifies that the neutral should never be grounded at any point on the load side of the service entrance on either single-phase or three-phase systems.

Figure 5.3 illustrates two correctly wired branch circuits. In Fig. 5.3(a), two pieces of equipment are energized from separate branch circuits. Minimum power-related voltages exist between the equipment cabinets (a) and (b). Thus, potential interference currents in the interconnecting ground path (provided, for example, by a ground bus, cable shield, etc.) are small. Figure 5.3(b) shows both pieces of equipment sharing a common-branch circuit, as in the same room, for example.

Figure 5.4 shows the effects of improper wiring. The condition of Fig. 5.4(a) is that of reversal of the *hot* (black) and *neutral* (white) conductors. Although a violation of the NEC, it does not

Sec. 5.3   Fault Protection Design

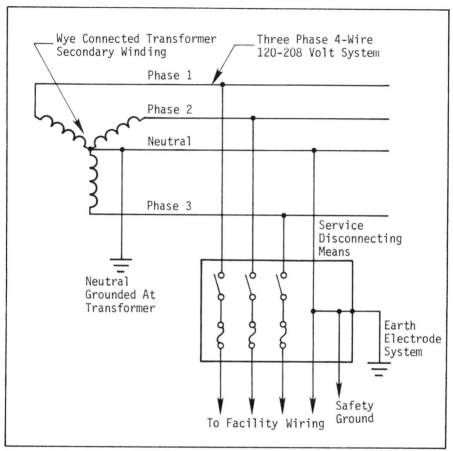

Figure 5.2 - Three Phase 120/208 Volt AC Power System Ground Connections.

automatically produce stray or return currents in the safety wire (or equipment cabinet/conduit system). The conditions of Figs. 5.4(b) and 5.4(c) represent the most likely problems and are the most troublesome from the standpoint of stray noise. Interchange of the neutral and ground conductors frequently exists because no short circuit is produced and the condition may go undetected. The full load current of terminating equipment however returns through the safety wire/conduit/cabinet ground system. The resultant common-mode voltages and currents pose a severe threat to interconnecting circuits between equipment elements (a) and (b). (The condition of Fig. 5.4(c) is the most troublesome from the standpoint of power frequency common-mode noise because of the higher voltage drop developed by the return current traveling through the longer path).

Sec. 5.3   Fault Protection Design

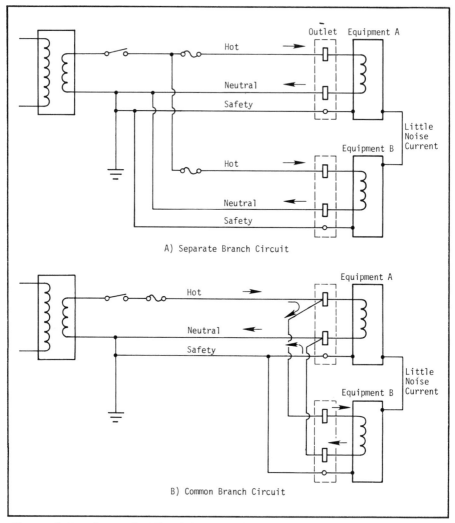

Figure 5.3 - Properly Wired AC Distribution Circuits for Minimum Ground Noise.

Sec. 5.3　　　　　　　　　　　　　　　　　　Fault Protection Design

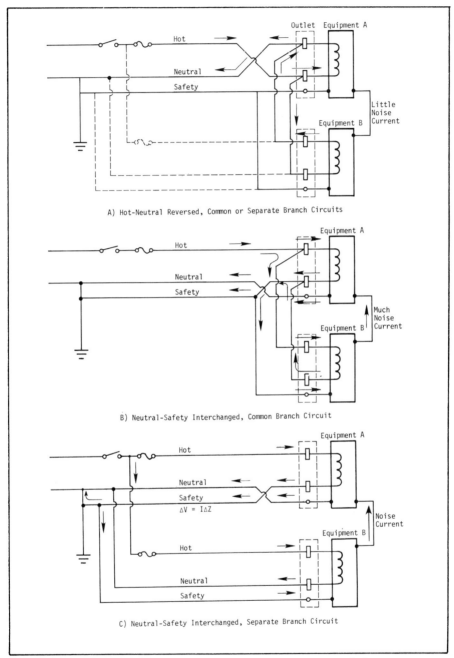

Figure 5.4 - Noise Problems Resulting from Improper Wiring.

Sec. 5.3   Fault Protection Design

Conditions 5.4(b) and 5.4(c) are the sources of many facility ground noise problems. Improperly grounded neutrals can occur in a piece of terminating equipment or can be the result of improper wiring at an outlet, junction box, or switch panel. (The most common cause is improper wiring). If the neutral is grounded at any point except at the service disconnect, part, if not most, of the load current returns from the load back through conduit, raceways, equipment cables (including coax and other cable shields), and structural support members. Voltage drops associated with this return load current appear as common-mode noise sources between separately located equipment, as illustrated. If this equipment must be interconnected with signal paths, appropriate common-mode rejection measures must be employed in the signal paths.

5.4 References

1. Dalziel, C.F., *Electric Shock Hazard*, IEEE Spectrum, Volume 9, No. 2, February 1972, pp. 41-50.

2. *Standard General Requirements for Electronic Equipment*, Requirement 1, MIL-STD-454C, 15 October 1970.

CHAPTER 6

System Grounding

The various grounding networks - lightning, power, and/or signal-are integral parts of a facility or system whether they involve vehicles such as automobiles, ships, missiles or aircraft; buildings, or building complexes. Like other facility components, grounding networks are exposed to the ambient electromagnetic environment. This environment includes many sources ranging from incident RF signals to lightning, and from atmospheric noise to internally generated stray currents. Because of the wide frequency range of signals in the environment*, the complexity of (collective) grounding networks and the traditionally suggested *Rules of Thumb* for grounding, system designers are often faced with a dilemma. The best way to avoid this dilemma is to consider the grounding requirements collectively. Also be prepared to make tradeoffs between circuit parameters, equipment shielding, system configuration, signal interfacing, and signal grounding while still meeting the electrical safety, lightning, and other requirements.

The objective and plan of the *facility ground network*** design should consider all aspects to the degree required for the particular mission of the facility in its environment and in the context of cost-effectiveness. One should attempt to integrate the three features of lightning protection, electrical safety, and EMI control.

---

* In a well shielded location such as inside of a ship, missile, or airplane, it *may* be acceptable to assume that the external environment is sufficiently attenuated to ignore all signals in the design of the grounding networks except those internal and intrinsic to normal system operation. However, the growing use of composite materials will likely soon disallow this assumption.

** The use of the term *facility ground network* is intended to be in the collective sense; it includes the power fault protection network, lightning protection network, building or structure (or frame, hull, fuselage skin, etc.) ground, equipment ground, signal ground, instrument ground, data ground, etc.

System Grounding

The integration of these features in the design may be difficult because they usually do not fall under the responsibility of the same individual, group, or perhaps a company, and they may not even be considered in the same time frame. For example, (in the case of a building) lightning protection and power grounding are often decided by an architectual and engineering firm during the structural design phase, and the EMI control aspects are left to the electronic equipment engineers for resolution long after the building is finished. When this happens, tradeoffs become difficult. However, the various grounding networks must interface and they all must operate in a common EM environment. To alleviate this situation in future facilities, equipment designers and EMI engineers must be allowed to participate early in the facility design process. (Aerospace systems are generally much better in this regard). For existing structures, those persons responsible for electronic equipment grounding should become thoroughly familiar with the power and lightning-protection ground networks through reviews of drawings, performance of current and voltage measurements, and physical inspections before deciding upon the final signal/equipment grounding networks.

6.1 The Designer's Dilemma

The following discussion presents a perspective of system involving land-based structures. However, analogous situations exist for vehicles, ships, planes, missiles and satellites.

The National Electrical Code imposes a requirement for a connection to earth at the service disconnect, a grounded neutral on single phase 120/220V circuits, and a grounding conductor run with the power conductors. Permanently wired equipment must have exposed metallic surfaces connected to the grounded conductor as dipicted by Fig. 6.1. Underwriters' Laboratory (UL) requires that cord-connected equipment have the third wire grounding conductor terminated to the case. Thus, as soon as the equipment (unless double insulated) is connected to the power source, the case is grounded. Consequently, any signal which is grounded or referenced to the case or cabinet, either directly or indirectly, automatically becomes grounded to the power distribution system (generally, power ground also means facility or structure ground).

Next, consider a lightning protection network. Both the National Lightning Protection Code[1] (ANSI C5.1) and the U_L Master Label[2] requirements specify that lightning conductors be crossbonded to metallic structural members to prevent flashover. The lightning ground should also be connected to the power ground and the utility pipes. When these requirements are met, then equipment (i.e., circuit) grounds will become interconnected with the lightning ground indirectly through the structural members and/or through the ac third wire grounding conductor. Therefore, metal-enclosed equipment is immediately exposed to the noise voltages on the power and building ground networks. When the equipment case is not grounded, the National Electrical Code is violated and the likelihood of lightning flashover to the case or to connected cables is increased.

How does one get out of this dilemma? Perhaps an alternative is to insulate instead of ground. Since equipment does not have to be (electrically) grounded if double insulated, could the electronic equipment case be insulated and thereby remove the need for the ac ground connection? Every control shaft, cable, and cable connector would then have to have this insulation provided throughout the system. The impracticality of this approach immediately becomes obvious. Further, the lightning flashover problem still exists. Wherever the equipment or the interconnecting cables are close to structural members (probably nearly everywhere), a flashover threat exists.[3] Cables or circuits extending outside structural boundaries would be particularly vulnerable because in the event of a strike, to either the structure or the extended conductor, extreme voltage differentials would be developed between the conductor and building ground.[4] High amplitude voltage surges would also be experienced at equipment terminals and across internal components. Consequently, insulation (or isolation, i.e., floating ground) is probably usable

6.3

Sec. 6.1 The Designer's Dilemma

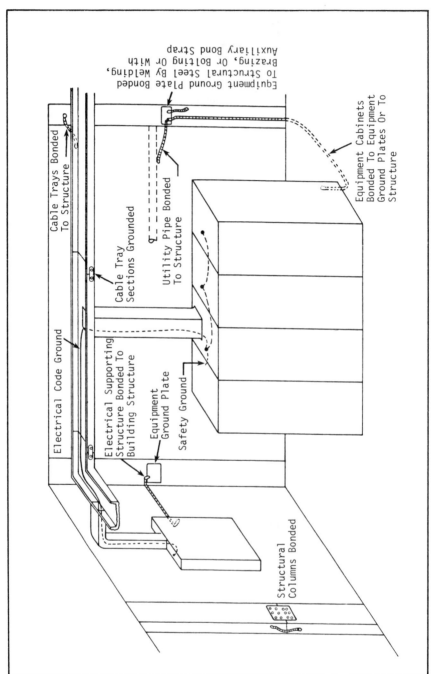

Figure 6.1 - Elements of a Facility Ground System.

Sec. 6.1 The Designer's Dilemma

only in very restricted circumstances. Therefore, it is likely that equipment cases and any signal circuits referenced to the case will have to be simultaneously grounded with the power, lightning protection, and building ground networks. Then what recourse does an equipment designer/installer/user have available for assuring that interference-free operation is obtained? The common advice is to *install a single-point ground* or *use multiple-point grounding*. Unfortunately, efforts to implement a tree grounding network* that is separate from the safety ground for equipment cases immediately run into conflict with the NEC. Generally, individual pieces of equipment interface and interconnect with other pieces across a room, in another part of the building, or outside. No realistic method exists for implementing a tree ground for equipment cases in a dispersed installation if (for no other reason) the various equipment units receive electric power from different locations on the ac distribution system. (The increased lightning hazard has already been noted).

The seemingly opposite suggestion is then to ground frequently, and with large, short ground conductors. The validity of this advice is heavily dependent upon the configuration of the system, the operating parameters (sensitivity, frequency range, impedance, etc.) of the equipment and on the nature of the EM environment. For example, a highly sensitive, high impedance, single ended, subaudio system will be susceptible to power frequency noise currents and voltages. This type of system will probably encounter considerable difficulty with noise if multiple-grounded because signal returns will be sharing, in effect, the ground-network conductors with all the stray environmental currents and voltages. An RF system, however, that uses constant impedance transmission lines for interfacing inherently operates single ended, i.e., outer conductor grounded to equipment case. Attempting to isolate such conductors from structural contact proves to be an almost impossible task, thus multiple-point grounding is the most practical procedure for such systems.[5,6]

The answer to the grounding dilemma will be found through the recognition that grounding is an overall system consideration, and must be approached as such. Tradeoffs in design, as with any complex interacting system, must be made.

---

* Ways for implementing a single-point (tree) ground for those systems requiring one are discussed in Sec. 6.4.

6.2 Basic System Grounding Requirements

The term *system* encompasses what the user wishes. For example, a *system* may represent a small self-contained electronic circuit totally within the confines of a case, cabinet, rack, or it may be an extended collection of equipment racks or consoles distributed over a wide geographical area. (The grounding requirements and procedures will be markedly different for the two different types of *systems*).

One way of distinguishing between different types of systems is to examine the manner by which power is obtained and how the equipment elements are interconnected with each other and with other systems. Based on these two considerations, systems may be identified as:

- Isolated
- Clustered
- Distributed
- Multiple-Distributed
- Central with Extensions

The properties of these types and the associated grounding requirements are presented in the following sections.

6.2.1 Isolated System

An *isolated** system is one in which all functions are accomplished with one equipment enclosure.

6.2.1.1 Characteristics

Only a single power source is associated with an isolated system. (Single power source means one battery pack, one branch circuit supply, etc.). In addition, only one ground connection (to structure, to earth, to hull, etc.) for the entire system is needed for personnel protection or lightning protection, or no facility ground connection at all is required. No conductors except the power cord and the appropriate ground exists because no interfaces (power or signal), with other equipment or devices which are

---

* An *isolated* system should not be confused with a floating system which has no external ground. Some isolated systems may be floating while others must be grounded.

Sec. 6.2   Basic System Grounding Requirements

grounded, are present or needed. Common examples of isolated systems are handheld calculators, desktop computers (off line), home-type radios, television receivers, etc.

6.2.1.2 Grounding of Isolated Systems

Grounding requirements for isolated systems are illustrated by Figs. 6.2 and 6.3. They typically include a third wire ground (unless the equipment is double insulated) if powered with single phase ac or a ground-cable run with the power conductors if multiphase is involved. Battery driven systems usually do not require any grounding (but be sure the system is isolated and not the *Central-with-Extension*---see Sec. 6.3.5). It is important that isolated systems are not located near lightning downconductors, or near other grounded metal objects in very high RF field strength environments (because of flashover or arcing). In certain very high-level RF environments, such as a broadcast transmitter, supplemental case grounds to prevent hazardous RF voltages from appearing on equipment cases may be necessary. The internal signal grounding requirements of the isolated system are those determined by the designer as necessary for proper self operation.

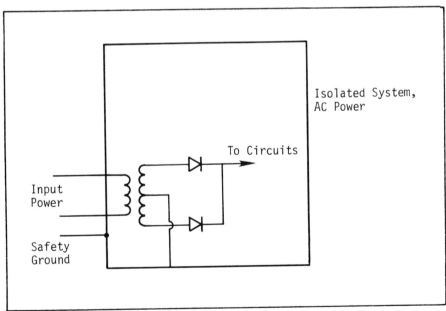

Figure 6.2 - Minimum Grounding Requirements for an AC Powered Isolated System.

Sec. 6.2   Basic System Grounding Requirements

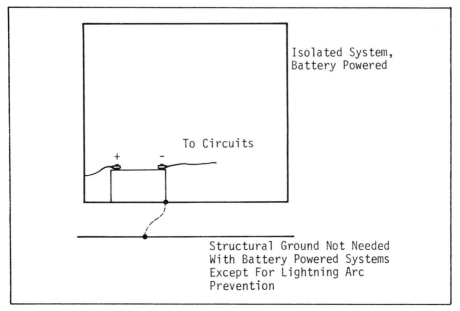

Figure 6.3 - Minimum Grounding Requirements for Battery Powered Isolated System.

6.2.2 Clustered System

Figure 6.4 illustrates a *clustered* system. Such a system is characterized as having multiple elements (equipment racks or consoles) located in a central area.[7,8,9] Typical clustered systems include minicomputers, component stereo systems, medium scale data processors, and multielement word processors.

6.2.2.1 Characteristics

A distinguishing feature of a clustered system is that it utilizes one common power source, e.g., a battery or a single ac power connection. There are likely to be multiple interconnecting cables (signal, control and power) between the members of the system but not with any other system. A clustered system only needs one facility (structure) ground tie to realize personnel safety and lightning protection requirements.

6.8

Sec. 6.2   Basic System Grounding Requirements

Figure 6.4 - A Clustered System.

6.2.2.2 Grounding of a Clustered System

Grounding for a clustered system requires that one connection be made to structural ground as illustrated by Fig. 6.5. If the power supplied is single phase ac, the third wire ground provides this connection. If the power supply is three phase ac, an apropriately sized (according to Para. 250-95 in the National Electrical Code) supplemental ground conductor is used. Battery powered systems should have one ground connection to the structure as shown in Fig. 6.6. The signal ground referencing scheme used between the elements of the system should reflect the particular signal characteristics (frequency, amplitude, etc.) of the various pieces of equipment. This scheme may be single-point* or multiple-point. If multiple point, the signal ground may be realized with cable shields, with auxiliary conductors, or it may utilize a wire grid or metal sheet under or above the array of equipment. In a benign (quiet) EM environment, signal grounding with cable shields or auxiliary conductors between the interconnected pieces is acceptable. This method of grounding a clustered system is depicted in Fig. 6.7(a). However, in high level, multisignal environments, this type of grounding scheme should be avoided because of the antenna pickup

---

* A more detailed discussion of single-point grounding is in Sec. 6.4.

6.9

Sec. 6.2 Basic System Grounding Requirements

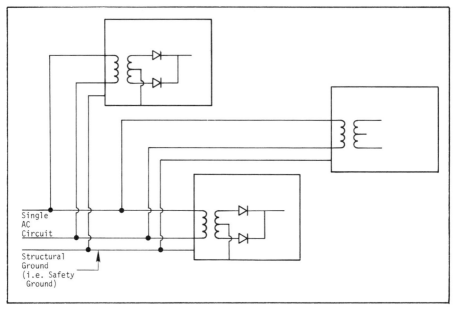

Figure 6.5 - Basic Grounding of a Clustered System.

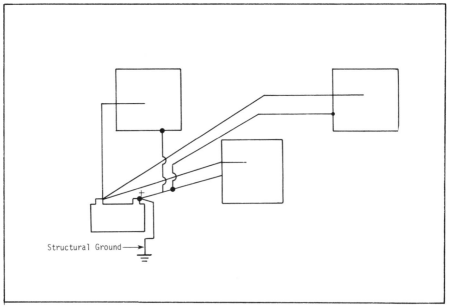

Figure 6.6 - Grounding of a Common Battery Clustered System.

Sec. 6.2 Basic System Grounding Requirements

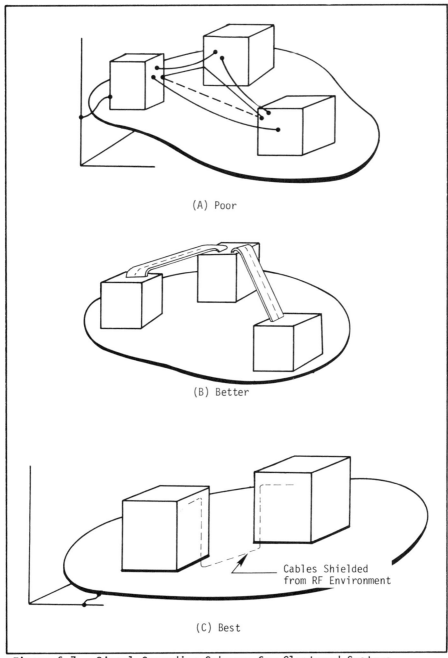

(A) Poor

(B) Better

(C) Best

Figure 6.7 - Signal Grounding Schemes for Clustered Systems.

Sec. 6.2 Basic System Grounding Requirements

effects of the multiple conductors. Signals coupling to the ground conductors produce common-mode voltages between various source-load pairs and raise the threat of interference. A better method of grounding is shown in Fig. 6.7(b). Here broad metal paths are provided between the pieces of equipment of the system. Overhead or underfloor ductwork, cable trays, and wire channels can frequently provide the necessary signal grounding and fault protection network for a clustered system. The best approach, and the one recommended for use in high level RF environments, is that shown in Fig. 6.7(c). A close-mesh wire grid or a solid metal sheet is provided for mounting the various pieces of equipment of the system. (In extremely severe environments as in the vicinity of a transmitter, the solid sheet is preferred over the grid). Each cabinet is carefully bonded to this grid or sheet. All interconnecting signal leads, power buses, etc., should be routed inside enclosures and beneath the ground plane, preferably in conduit or in raceways. Note that the evolution from the poorest to the best approach is aimed at rendering the system and its cables successively less effective as a pickup *antenna* for radiated RF energy, particularly in the broadcast, HF, VHF and UHF bands.

6.2.3 Distributed System

A distributed system is one in which major elements are physically separated in a way that requires equipment to be variously fed from different power outlets, branch circuits, different phases of the line, or perhaps even different transformer banks.

6.2.3.1 Characteristics

In a distributed system, separate safety and lightning protection grounds to facility (vehicle) structure (frame) are required. Another common characteristic of a distributed system is that multiple conductor (signal and control) paths exist between system elements. Conductor lengths are likely to be greater than $\lambda/10$ at frequencies where an interference threat exists. Examples of distributed systems include industrial process control, environmental monitoring and control, communications switching, and large main frame computer nets.

6.2.3.2 Grounding of Distributed Systems

Effective grounding of distributed systems to achieve the required safety and lightning protection for equipment and personnel, while minimizing noise and EMI, requires careful application of the principles set forth in Chapters 2, 3, and 4. To describe a

Sec. 6.2 Basic System Grounding Requirements

stereotyped network or to list a set of rules is not considered prudent. However, some general guidelines for proceeding may be suggested.

If the system is ac powered, consider each major element (consisting of one or more types of equipment essentially located at a particular location) as either an isolated or clustered system, as appropriate. Proceed to ground each major element as discussed in Secs. 6.2.1.2 and 6.2.2.2. Each and every signal port on these *isolated* or *clustered* subsystems that must interface with other portions of the total system, i.e., other *isolated* and *clustered* subsystems, should be viewed as having to interface with a noisy world. As such, the techniques discussed in Chapter 3 for controlling unwanted coupling of radiated and conducted interference into the signal paths must be fully employed. Obviously, discretion will be necessary. There will be situations, depending upon the properties of the signal being transmitted from terminal to terminal, the characteristics of the signal path, and the nature of the electromagnetic environment, in which no additional protective measures are required. In general, one must insure that adequate isolation is provided between the external (to the system) conducted and radiated noise envrioment and the signal path.

Common battery distributed systems present a particular challenge. (Such systems are commonly found on aircraft where the fuselage is used for dc power return). The use of the equipment rack, structure, hull or fuselage for the power return path means that the structure becomes the circuit reference. Therefore, voltage differentials between various points in the structure appear in series with any single ended, unbalanced signal paths. Again, adequate rejection against such conducted interference must be obtained through application of the various techniques discussed in Chapter 3.

6.2.4 Multiple Distributed Systems

As the name implies, the characteristics of multiple distributed systems are similar to that of a distributed system except that there are usually several systems contained and operating in the same general area. Typically, the multiple systems share the primary power sources. A distinguishing feature of multiple distributed systems is that they typically run a high risk of interfering with each other and are susceptible to interference from facility noise and the external environment. Thus, in addition to having grounding requirements like those discussed for distributed systems, additional shielding and filtering requirements are necessary to minimize intersystem interference.

Sec. 6.2   Basic System Grounding Requirements

6.2.5 Central System with Extensions

This type of system is illustrated in Fig. 6.8. It is distinguished from an isolated or clustered system in that integral elements of the system extend out from the central portion at long physical and electrical distances.

6.2.5.1 Characteristics

This system is distinguished from a distributed system in that the extended elements obtain their power from the central element. Connections to a power source are not made anywhere except at the main element. An example of this type of system is an industrial process controller with sensors and actuators located remotely from the data logger or controller.

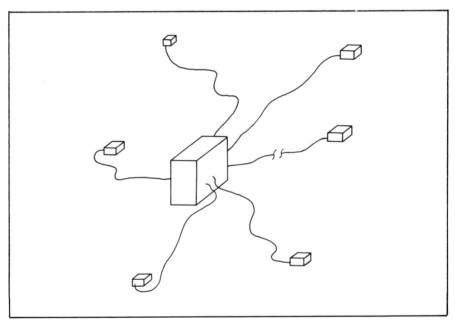

Figure 6.8 - Central-With-Extensions System.

6.2.5.2 Grounding of a Central System with Extensions

The central or primary element of this type of system should be grounded as though it were a clustered or isolated system. Depending upon the operating frequency ranges and signal levels of the extension elements and the characteristics of the EM environment, a single-point tree or star grounding scheme may be used or a multiple-point scheme may be used. It is likely that most systems of the central-with-extensions configuration will involve relatively low frequency (audio or below) with operating bandwidths encompassing the power frequencies. If this is the situation, a single-point tree is recommended. The ground mode would be at the central element with one connection (the safety ground) made to the structure. Extended elements should be floated or balanced. Twisted pair or balanced signal transmission-line conductors should be used between the central element and the extended elements. If radiated coupling proves to be a problem, the extended elements should be configured, if at all possible, so that the shields (as on coax, for example) from the central element can be continued to enclose the extended elements.

6.3 Single-Point Grounding

Structural members, utility pipes, and electrical supporting structures (i.e., the conduit, raceways, etc.) often provide a very noisy reference plane for signal circuits. This electrical noise results from stray power currents, from lightning and from inductive coupling to power lines and lower frequency RF sources. Many types of electronic systems are potentially very susceptible to these low frequency noise currents and voltages that are present in the structural *ground plane*. If both the source and load end of a potentially susceptible source pair are both connected to this ground, common-mode interference is likely. An obvious approach to the avoidance of this type of interference situation is to open the *ground loop* by connecting only one end of the signal transfer path to ground. The end to be grounded should be the one that maximizes the signl-to-noise ratio in the loop--generally the source end. There will be situations (multiple sensors are feeding one processor, for example) where grounding every *source end* will create a multiple-grounded system and thus the load (e.g., the central processor) must be the ground point in order to maintain a single-point ground.

Where there are multiple elements to a system, where the ground plane is noisy from conducted or induced signals, and where the signal circuits are susceptible to the ground-plane noise, single-point grounds *from the circuit point of view* may be required. For example, consider a circuit/equipment pair like that illustrated in Fig. 6.9. A typical design situation is depicted where a transformer-type power supply feeds internal circuitry. Power-line filter capacitors on the transformer primary are shown to illustrate leakage paths for stray currents. The secondary side of the transformer is referenced to circuit ground which is connected to equipment ground which, in turn, is connected to the building ground. Equipment on the receiving end is configured similarly. Note that when each equipment cabinet is connected to different points on a noisy building ground, noise voltage V_n appears directly in series with the desired signal.

The preferred way of eliminating the effects of V_n is illustrated in Fig. 6.10. Note that for an individual piece of equipment (serving as either a source or a load for a signal) only one connection is made to the reference ground.

Figure 6.11 shows an alternative approach to *single-point* grounding. One can see that circuit grounds inside equipment boxes are not connected to equipment cabinets. Instead, they are grounded with isolated cables that are connected to the structural ground reference at one common point. This approach is effective where the noise threat arises from conducted noise, from electric fields, and from plane-wave fields. However, it should not be used in magnetic field environments (such as that associated with nuclear electromagnetic pulses) because of the loop that would be formed by the

Sec. 6.3   Single-Point Grounding

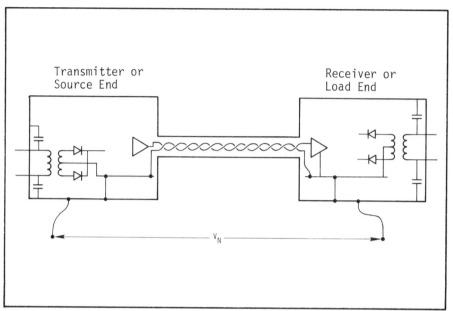

Figure 6.9 - Conducted Noise Threat.

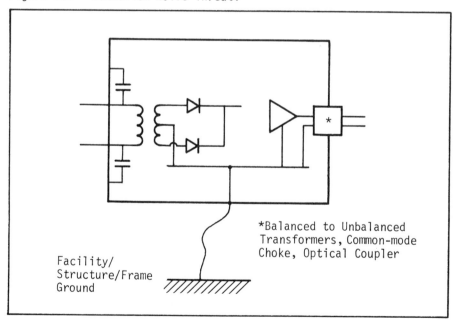

*Balanced to Unbalanced Transformers, Common-mode Choke, Optical Coupler

Figure 6.10 - Preferred Grounding of Low Frequency Equipment in Noisy Environments.

Sec. 6.3   Single-Point Grounding

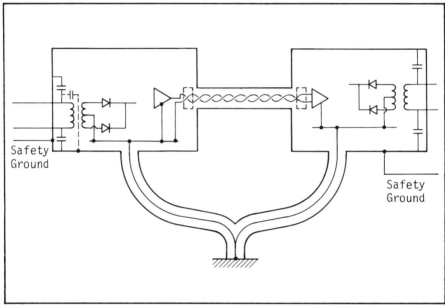

Figure 6.11 - Alternative No. 1 for Single-Point Grounding.

signal return and the ground conductors.

Figure 6.12 illustrates another less effective method of implementing the approach of Fig. 6.11. Since the grounding conductors are exposed to all types of radiated coupling, they are likely to serve as very effective antennas and develop severe noise differentials between the equipment boxes. Consequently, this approach should be used very sparingly and only in those circumstances where it is known that radiated (both near field and plane) pickup is extremely limited or if the resulting noise differentials present no problems to the operating circuits.

Single-point grounding systems have been recommended for certain specialized facilities where particular operational needs are known to to exist.[10] Before adopting such approaches for other facilities, the preceeding precautions must be considered. In particular, it should be fully recognized that the approaches of Figs. 6.11 and 6.12 do not function like those illustrated in Fig. 6.10.

In summary, single-point grounding has its areas of effective application. For greatest effectiveness, however, it must be a true *single-point* ground and not a mesh or loop with a single connection to structure or power ground. It should not be expected to circumvent the need for other common-mode noise control measures. It works best when combined with those measures.

6.18

Sec. 6.3   Single-Point Grounding

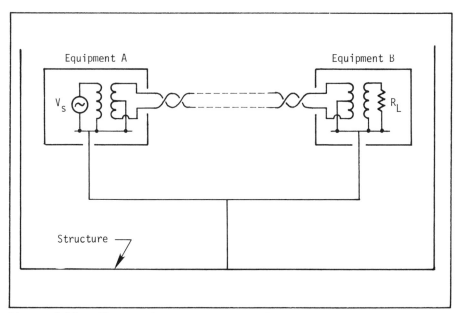

Figure 6.12 - Alternative No. 2 for Single-Point Grounding.

6.4 Conclusions

The grounding of complex systems to minimize EMI, while meeting other requirements for safety and lightning protection, is a complex task. Rules are difficult to formulate because all systems tend to have different operating characteristics, are differently configured, and are exposed to different environments, etc. The first step is to define the type of system under consideration. Then, depending upon the type of system, certain recommended steps can be followed. The basic principles involved in the recommendations include minimizing the amount of common-mode exposure to the various system elements and implementing steps to provide maximum discrimination (or rejection) to the residual common-mode noise threat.

6.5 References

1. *Lightning Protection Code 1968, NEPA 78*, National Fire Protection Association, Boston, Ma.

2. *Master Labeled Lightning Protection Systems*, UL 96A, Underwriters' Laboratories, Inc., Chicago, Il, June 1963.

3. Smith, R.S., *Lightning Protection for Facilities Housing Electronic Equipment*, FAA/FIT Workship on Grounding and Lightning Protection, FAA Report No. FAA-RD-77-84, Melbourne, Fl., May 1977, pp. 267-282.

4. Bodle, D., *Electrical Protection Guide for Land-Based Radio Facilities*, Joslyn Electronic Systems, Santa Barbara Research Park, P.O. Box 817, Goleta, Ca 93017, 1971.

5. LaDieu, F., *Frequency Division Multiplex Baseband Cable Plant Improvements*, FAA/GIT Workshop on Grounding and Lightning Protection, FAA Report No. FAA-RD-78-83, Atlanta, Ga, May 1978, pp. 35-92, AD A058-797.

6. Turesin, V.M., *Theoretical Analysis and Design Techniques for Grounding to Accomplish EMI Control*, FAA/FIT Workshop on Grounding and Lightning Protection, FAA Report No. FAA-RD-77-84, Melbourne, Fl, May 1977, pp. 323-330.

7. Condon, G.P., *A System Grounding Approach for High Speed Digital Computer Generated Image (CGI) Simulation Devices/Trainers*, FAA/FIT Workshop on Grounding and Lightning Protection, FAA Report No. FAA-RD-77-84, Melbourne, Fl, May 1977, pp. 441-462.

Sec. 6.5 References

8. Gowdeski, G.W., *On the Design of Chassis Ground Networks Used in Large Scale Digital Simulator Systems*, FAA/FIT Workship on Grounding and Lightning Protection, FAA Report No. FAA-RD-77-84, Melbourne, Fl, May 1977, pp. 463-474.

9. Hokkanen, R.N., *Design Guide for the Utilization of Raised (Computer) Floor Systems as an Integral Part of the Grounding System*, FAA/FIT Workshop on Grounding and Lightning Protection, FAA Report No. FAA-RD-77-84, Melborune, Fl, May 1977, pp. 475-506.

10. Denny, H.W., et al, *Grounding, Bonding, and Shielding Practices and Procedures for Electronic Equipment and Facilities*, Report No. FAA-RD-75-215, I, II, III (3 Volumes), Contract No. DOT-FA72WA-2850, Engineering Experiment Station, Georgia Institute of Technology, Atlanta, Ga, December 1975.

CHAPTER 7

Equipment Grounding

Electronic equipment is fundamentally a collection of electronic circuits which generate, detect, process or act on signals or inputs in some prescribed manner. As noted in Chapter 2, these various circuits are basically interconnected source-load pairs. All of the active electrical and electronic circuits in a piece of equipment can be reduced to some combination (ranging from very simple to very complex) of sources and loads and their conductor paths.

The purpose for grounding equipment is to realize the signal, power, and electrical safety paths necessary for effective performance. Grounding from the EMI standpoint can be viewed as (1) the realization of these functions without introducing excessive common-mode noise and (2) establishment of a path to divert interference energy on external conductors, and present in the environment, from entering susceptible circuits or components. Equipment grounding simply means the application of the principles and techniques at the equipment and circuit level that were set forth in the preceding chapters. Many grounding paths at the device and circuit level are extremely short and therefore grounding often becomes synonomous with bonding, the subject of the suceeding chapter.

7.1 Circuit Return Coupling

Every conductor exhibits properties of resistance and inductance, and between every pair of conductors there is capacitance. This situation applies equally as well to extensive sheets of metal and printed-circuit board conductors as it does to wires. Thus, every signal, power and control common (or return or ground) will offer some impedance to the currents flowing in it. The resulting voltage poses a threat of interference or coupling to other circuit pairs. The relative magnitude of the interference signals, as noted earlier, is dependent upon the impedance of the common path (which is a function of material and geometry of the conductors and the frequency of the signals), the amplitude of the offending signals, and the sensitivity of the susceptible circuits. Control of this unwanted coupling follows the principles described in Chapter 3 whether the resulting problem is excessive feedback resulting in oscillations, power rectifier switching transients causing hum, or cross talk from one circuit to another.

Grounding is harder to visualize at the circuit level than where long cable runs are present. However, consider that at the point where current returns connect to ground the voltage drop in the ground is low. Voltage contours exist at increasing physical distances from this point because of the impedance properties of the path. Figure 7.1 illustrates the nature of these contours. These voltage contours represent a common-mode noise source which can cause interference coupling or circuit instabilities.

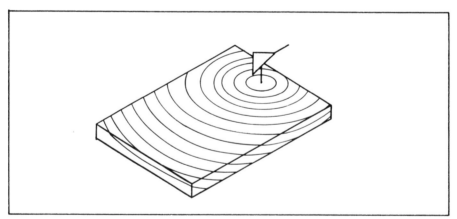

Figure 7.1 - Chassis Voltage Contours Produced by Ground Circuits.

7.2 Low-Frequency Circuit Grounding[1]

Amplifier feedback resulting from ground-plane coupling is shown in Fig. 7.2. Heavy currents from the output amplifier returning to the power supply produce voltage drops in the chassis (or the pc board return bus). The contours as depicted represent a voltage drop in series between the input jack ground and the first stage amplifier ground . Such a drop shows up across the first stage input and is amplified. Depending upon the phase relationship through the amplifier chain, the coupled signal may cause distortion, oscillations, or other problems. Some immediate solutions to the feedback problem of Fig. 7.2 are:

- Locate the final power output stage as close to the power supply as possible to minimize the voltage drop through the supply and return leads.

- Physically isolate the input amplifier and the input jack from the chassis and the power output ground.

- Balance the input amplifier, the input connector, and the intervening conductor. Very high gain amplifiers may require the use of shielded input cables and guarded amplifiers (see Sec. 7.6).

- Use optical isolators between the input jack and the preamplifier.

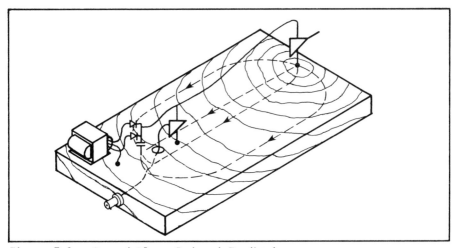

Figure 7.2 - Ground Plane Induced Feedback.

Sec. 7.2 Low-Frequency Circuit Grounding

Figure 7.3 illustrates the interstage coupling that occurs between the output and input as a result of power supply and ground return drops. Assume that the amplifier chain is constructed on a printed circuit board with the supply, input and output leads passing through a multipin connector. If the chain is physically laid out like the schematic shows, that is the supply leads and ground are connected first to the lowest level stage and then sequentially to the highest level stage, ample opportunity exists for coupling to occur. The problem becomes even more severe if the input signal is referenced to *ground* at the connector.

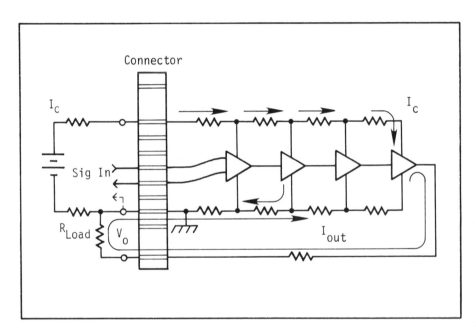

Figure 7.3 - Interstage Coupling Caused by Improper Grounding.

A much better layout is depicted by Fig. 7.4. Observe in particular that the power supply and return leads are run directly to the high level output stage. The return may or may not be coincident with the board's foil common. In many cases, it can be, but it depends mainly on what other circuits must use the foil common and whether or not the voltage contours will cause coupling problems. In any case, a separate negative (low side) conductor should connect to each low-level stage and make only one point-of-contact with ground at the output amplifier ground connection. Similarly, the positive supply lead must be kept separate from the low-level stages except

Sec. 7.2 Low-Frequency Circuit Grounding

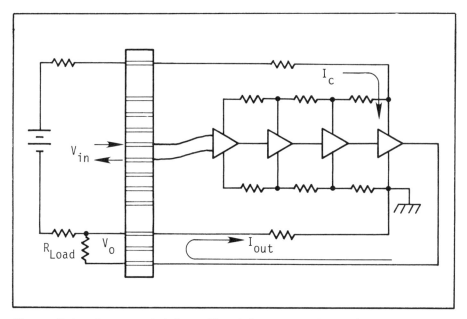

Figure 7.4 - Recommended Grounding Scheme for PC Board Mounted Chain Amplifiers.

at the connection to the output stage. In *all* cases, keep the low side of the input signal separate from the negative return (*ground*) pin of the connector. It is necessary that these concepts be extended to all elements of the system to obtain maximum benefit from such an ordered approach to circuit board grounding. Otherwise, the benefits will be lost because of compromises external to the board.

The same type of supply and return lead coupling contributes to interference between stages sharing the same circuit board or the same supply. These concepts for a chain amplifier can be extended to multiple circuits. For example, consider a collection of different circuits such as those shown in Fig. 7.5. To minimize coupling between such circuits, establish supply and ground reference nodes on the circuit board - preferably as physically close to the circuit element with the heaviest current demands as practical. Bring supply (both polarity) leads from the connector pins to the node points. Connect the supply and ground for the heaviest load, both in terms of heaviest current and greatest rate of change (dI/dt, dV/dt), to those nodes with minimum length conductors (wire or pc type). Attach adequate filter capacitors to the nodes, if at all possible, to minimize ripple on the supply leads. Run a separate, dedicated secondary supply and ground conductor from each self-compatible circuit, or assembly, to these node points as shown in Fig. 7.6. If this is not possible because of board layout restrictions or because of excessive

7.5

Sec. 7.2   Low-Frequency Circuit Grounding

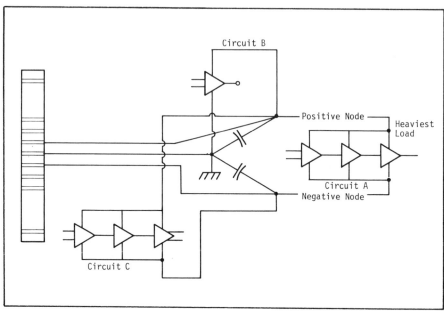

Figure 7.5 - Grounding of Potentially Incompatible Circuits Sharing a Common Circuit Board.

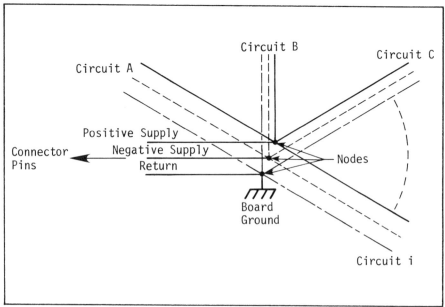

Figure 7.6 - The Star Ground.

Sec. 7.2   Low-Frequency Circuit Grounding

conductor lengths (important where high speed/high frequency signals are involved), share secondary supply conductors only between circuits with minimum likelihood of unwanted interaction. Any digital or other switching circuits should be laid out to place them close to the nodes -- the higher the data rates and the heavier the currents switched, the closer the circuits should be to the nodes. (Where very-high data rates are involved, conductors must be as wide as space will allow). Note that the described arrangement represents the often mentioned *star* or *single point* ground.

With an ideal single-point ground system, each stage is connected through its own independent ground path back to a single point (a node, root, etc.) somewhere in the system. In this way, the ground return or reference for any one stage contains the ground return current from no other stage. It is essential that the supply and ground nodes be as close to the power supply regulator as possible to achieve the full benefit of single-point grounding. In many cases, it may be necessary to provide secondary, supplemental regulation near indivdual circuits especially where one common supply serves many circuits. Circuit signal ground must be isolated from chassis everywhere except at the ground point or node. (If this isolation is compromised *anywhere*, the benefits of a single-point ground are lost).

A pure single-point ground down to the smallest identifiable circuit is not likely to be possible in a system of any degree of complexity. A reasonable compromise is to provide a single-point ground return and power supply to each plug-in assembly. Within each major chassis, each plug-in unit should be tied to the major chassis power supply and ground wires at the single ground for the assembly. If the chassis is mounted on a frame which may carry large currents, electrical contact between chassis and frame should occur at only one point. Circuits should be laid out to minimize the chassis area in which high frequency currents flow.

7.3 Power Supply Filter Grounding[2]

Power supply hum is often the result of careless grounding. For example, suppose that the power-supply filter is physically separated from the point where the transformer secondary centertap (for a full wave rectifier) is grounded so that the capacitor charging currents must return through the chassis or circuit board foil. If sensitive, high-gain input stages are grounded at one point to the chassis and the input terminals are grounded to a different point so that ground-plane voltage differentials are present between the two signal ground points, then noise or hum will appear in the circuit as illustrated by the circuit diagram of Fig. 7.7. Some solutions to this hum problem are set forth in the preceding section. In addition to those suggestions, assure that the low side of the power supply filter capacitor physically connects to ground at the same point as the transformer secondary or, in the case of a full wave bridge, at the same grounding point as the bridge rectifier as shown in Fig. 7.8.

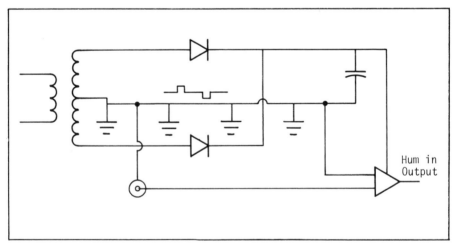

Figure 7.7 - Power Supply Hum Caused by Ground Return Coupling.

Sec. 7.3   Power Supply Filter Grounding

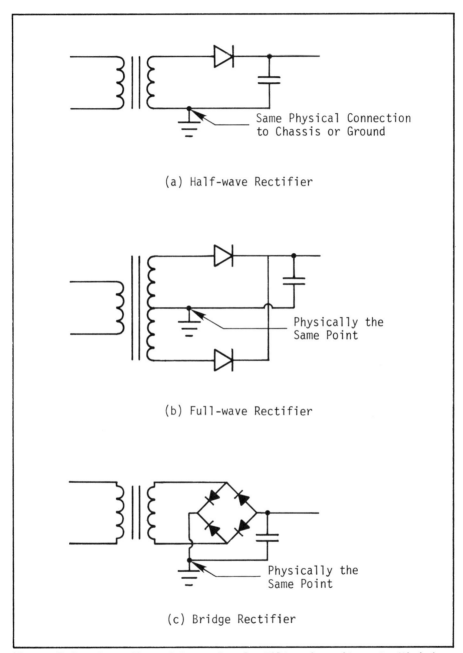

Figure 7.8 - Grounding of Power Supply Filter Capacitors to Minimize Ground Noise.

7.4 Digital Circuits

Various logic families exhibit different properties. All, however, possess some common characteristics that influence how they should be grounded. The signals traversing between the various elements (chips, boards and equipment) are primarily binary in nature: a *low* state defined as being below some voltage threshold (from a few tenths of a volt) and a *high* state defined as being above some voltage threshold (from one to 10 volts depending upon the family). The rise times of the pulses may range from microseconds to nanoseconds, requiring transmission bandwidths from a few megahertz to several hundred megahertz. The one common characteristic that strongly impacts circuit and equipment grounding is, since the logic elements typically will respond to dc voltages exceeding the thresholds, that the low end of the response bandwidth of a digital circuit is likely to extend to zero hertz (dc). Thus, if both the driver and receiver of a digital pair are both grounded to a noisy reference, common-mode noise usually produces data or computational errors. For this reason, many logic system designers seek to implement a single-point ground. This is normally done through isolating internal circuits from chassis and installing a signal or *logic* ground cable that is isolated from all chassis and structure except for one connection, as shown in Fig. 7.9. The advantage of this approach is that the structure and/or cabinet ground-plane noise is not directly coupled into the logic reference. The disadvantages are (1) the logic ground cables may act as unwanted pickup antennas for radiated interference in active EM environments and (2) damaging voltages may be produced between the cabinet and internal circuitry during lightning strikes or an EMP event unless adequate surge protection is provided.

Implementation aspects of a single-point ground system where signal ground is isolated from chassis are illustrated in Fig. 7.10. The important points to note are:

- The secondary power supply should *not* have its reference, i.e., centertap, connected to *chassis*. instead *all* power leads (+/- return) should be carried through the power supply compartment wall via properly filtered connections (that is, where the primary power supply lines have been treated to meet conducted and radiated emissions and susceptibility requirements, e.g., MIL-STD-461) or directly to the pc board elements as discussed earlier (see Fig. 7.11).

- All secondary power supply filters and regulators should be referenced to the returns and not to chassis.

Sec. 7.4   Digital Circuits

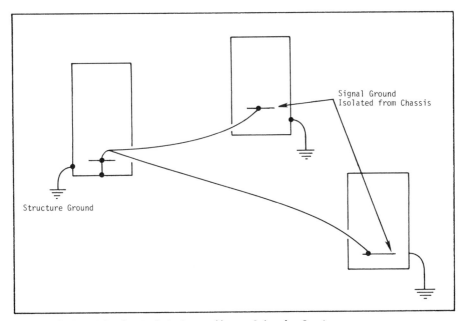

Figure 7.9 - Single Point Grounding of Logic Systems.

- The secondary supplies in all equipment in the system and sharing the one common ground should be handled in the same way.

- Any line drivers and receivers serving interconnecting data buses should be balanced and twisted-pair cables should be used where magnetic field coupling is a threat. Each wire of the twisted pair should be connected only to the output terminals of the line drivers and to the input terminals of the line receivers. Neither of the data lines should be connected to signal (pc board) common.

- Between interfacing equipment, dedicated wires may be used to interconnect pc (data) commons, if the system is intended to operate in this manner. Otherwise, specific steps must be taken to interface the equipments in a manner that does not violate the single-point grounding system. Such steps include the use of transformer couplings, common-mode chokes, optical isolators, frequency translation, etc.

Sec. 7.4 Digital Circuits

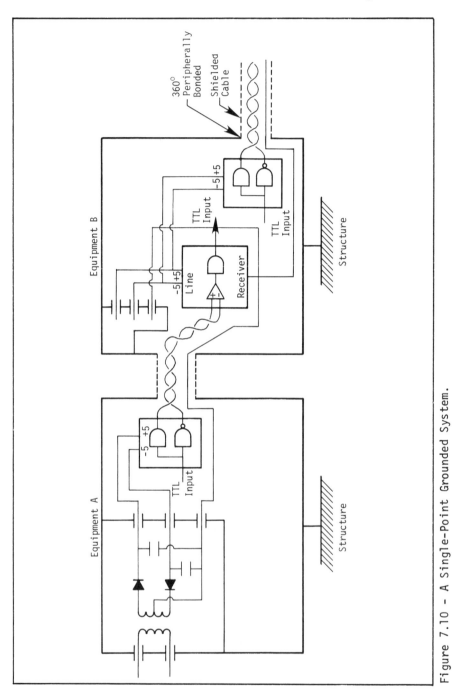

Figure 7.10 - A Single-Point Grounded System.

Sec. 7.4 Digital Circuits

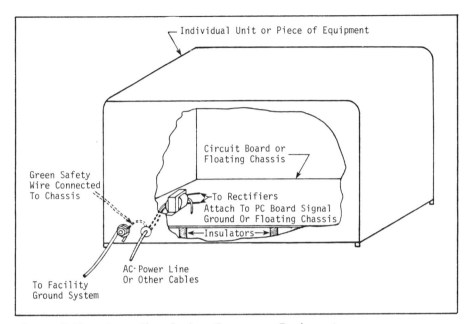

Figure 7.11 - Grounding in Low Frequency Equipment.

- In *one* piece of equipment, the pc common, and the secondary power supply return will be connected to the chassis. This connection is made internal to the equipment. (This requirement is particularly important in high level RF environments where pickup by the ground lead would be detrimental).

- The data transmission lines and the dedicated common lead should be included within the same cable shield, if used.

- Cable shields should be of a type that provides effective shielding over the complete frequency range of the operational environment. Both cable shield and equipment shield integrity must be maintained through proper cable shield termination practices, e.g., peripheral bonding of cable shields at exit and entry points.

- In high lightning areas, surge suppressors must be mounted between the pc board ground and the equipment chassis in all equipment other than the one where the signal common to chassis ground connection is made. Power-line arrestors should be referenced to chassis at the power cable entrance to the housing.

7.13

Sec. 7.4					Digital Circuits

Another approach often used with data processing equipment is illustrated in Fig. 7.12. With this approach, all equipment boxes except one are isolated (insulated) from structure. This approach cannot be recommended for several reasons. Practical experience shows that maintaining isolation between enclosures and structure and other grounded objects proves to be almost impossible over long periods of use. Arcing hazards between the isolated cabinets and structure and other grounded objects during lightning storms are likely to arise, particularly if the various equipment units are widely distributed throughout the facility. Lightning-produced surges on data lines are also likely to be acute. Finally, where the various elements of a system are of such a nature (or so located) to require distributed sources of ac power, it is extremely difficult to maintain necessary isolation while complying with electrical safety codes. For example, both the National Electrical Code and the Underwriters' Laboratories require that a safety ground wire be run with the power-supply conductors connected to the cabinet (enclosure) and the power-system neutral at the disconnect. Where separate power cables must supply separate elements of the system, requirements for the safety conductors prevent isolation from being realized.

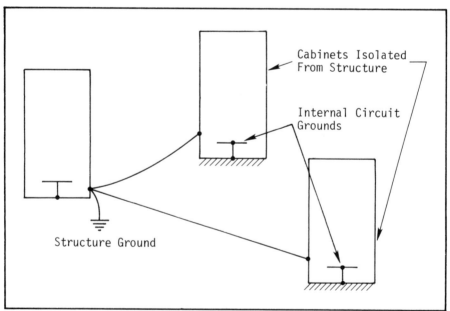

Figure 7.12 - An Alternate Single-Point Logic Ground System.

7.5 Instrumentation Grounding[3,4,5,6]

Many, if not most, data instrumentation systems are concerned with measurement or detection of physical phenomena (or changes in them) that require periods of observation or measurement that range from a few milliseconds to several minutes or hours. Because of the relatively slow nature of the event, the fundamental frequency of the transducer output may range from 0 (dc) to a few hundred Hz. Outputs of the transducer are commonly analog in nature. Power distribution systems, electromechanical switches, and atmospheric noise produce extraneous voltages whose energy content is strongly concentrated within this low-frequency region. Because of this overlap of signals, special techniques are generally required to keep voltages or currents produced by the extraneous sources from obscuring the transducer outputs.

Since transducer signals are primarily low frequency in nature, a basic single-point ground should be implemented. The signal return line should be grounded at one end only or not at all (i.e., it should be balanced). Similarly, individual cable shields around signal lines should be grounded at one end only.

7.5.1 Grounded Transducers

The bonded (grounded) thermocouple, illustrated in Fig. 7.13, is used with a single-ended data amplifier whose output drives recording devices such as oscillographs, strip-chart recorders, and magnetic tape recorders. Certain important aspects which should be considered are as follows:

- The shield which surrounds the transducer signal leads should be grounded at the same point as the transducer.

- When the bonded thermocouple is connected to an isolated differential amplifier as shown in Fig. 7.14, the shield of the input cable should be connected to the amplifier internal guard shield to continue the signal shield to within the amplifier. Notice that a grounding bus is shown connected between the data system signal reference and earth ground (structure) of the test area. This ground bus is necessary in any instrumentation system which uses isolated differential amplifiers. This is done to (1) provide the earth reference for the signal circuitry within the recording system to reduce high voltage hazards, and (2) minimize the common-mode potentials that otherwise exist between the

Sec. 7.5   Instrumentation Grounding

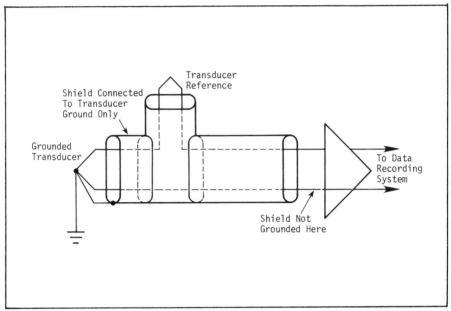

Figure 7.13 - Grounding Practices for Single-Ended Amplifiers.

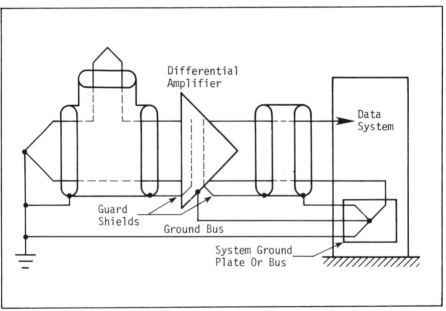

Figure 7.14 - Grounding Practices for Differential Amplifiers.

Sec. 7.5 Instrumentation Grounding

amplifier's input and output if the data recording system is grounded to a separate earth or facility ground. Notice that the amplifier case and output shield are connected to the data system (or load end) ground.

- Grounded bridge transducers should be excited with a balanced dc source. By balancing the dc excitation supply relative to ground, as shown in Fig. 7.15, the entire bridge will be balanced with respect to ground. Any unbalanced impedance presented to the amplifier input will be due to the leg resistances in the bridge. Although a ground loop still exists, its effect is greatly reduced by a balanced excitation supply.

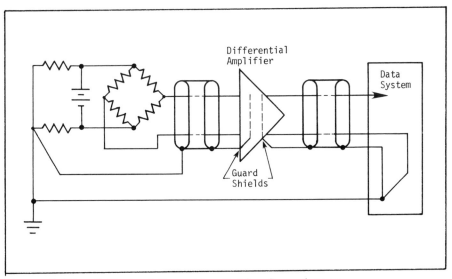

Figure 7.15 - Method of Grounding Bridge Transducers.

- Wherever possible, use an isolated amplifier with bridge transducers in the manner illustrated in Fig. 7.16. (With this configuration, both the transducer and the amplifier can be grounded without degrading system performance).

Sec. 7.5 Instrumentation Grounding

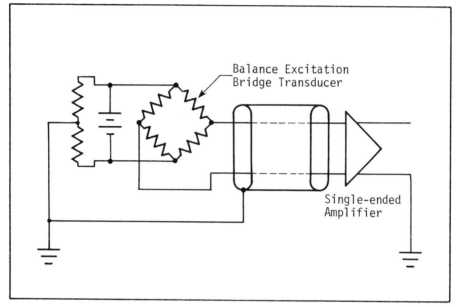

Figure 7.16 - Use of Isolated Differential Amplifier with Balanced Bridge Transducer.

- Provide a single common-signal ground reference point for all grounded transducers at the test area or on the test item.

- Connect the instrumentation cable shield of each data channel as close to the transducer ground connection as possible.

- Use twisted shielded transducer extension wires.

- Use a floating load on the output of a single-ended data amplifier when the amplifier input is a grounded transducer.

- Connect guard shield of data amplifier to input cable shield.

- Always use insulated shielded cables. Uninsulated shields should never be used in data instrumentation systems.

Sec. 7.5　　　　　　　　　　　　　　　　Instrumentation Grounding

7.5.2 Ungrounded Transducers

- Grounding techniques recommended for ungrounded transducers are illustrated in Fig. 7.17. The metallic enclosure of the transducer is connected to the cable shield and both the enclosure and the shield are grounded at the transducer. If the load on the cable signal line is a single-ended amplifier as shown in Fig. 7.17(a), the shield of the input cable should not be connected to the amplifier. The case of the amplifier should be grounded at the load.

- When using an isolated amplifier, the recommended method of grounding the system is shown in Fig. 7.17(b).

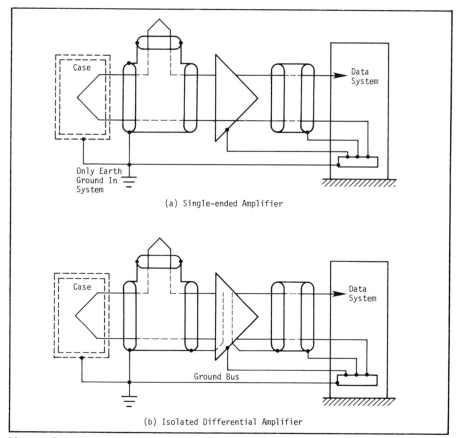

Figure 7.17 - Recommended Grounding Practices for Floating Transducers.

Sec. 7.5 Instrumentation Grounding

(Certain types of non-isolated differential amplifiers require that a transducer ground path be provided for proper amplifier operation. The instructions supplied by the amplifier manufacturer should be consulted for correct procedures).

- Provide a single common-ground reference point for all cable shields.

- Ground all input cable shields at the transducer.

- Provide a continuous overall shield for signal wires from transducer case to the input of the data amplifier.

- Connect isolated amplifier guard shield to input cable shield.

- Do not allow more than one ground connection in each input cable shield.

7.5.3 Transducer Amplifiers

- Single-ended amplifiers can be used in digital data acquisition systems if channel-to-channel isolation is provided (e.g., through the use of floating loads).

- Single-ended amplifiers should not be used with grounded (bonded) transducers (in order to avoid channel-to-channel ground loops).

- Single-ended amplifiers should not be used with grounded bridges (to avoid short circuiting one leg of the bridge).

- Connect amplifier output guard shield to data system ground bus.

- If a permanent unavoidable instrumentation ground exists at the test item, as well as at the data system, use isolated differential amplifiers to break the ground loop.

7.6 Construction Guidelines for Low Frequency Equipment

The signal ground network in low frequency equipment must be designed and installed to provide complete electrical isolation between the ground network and the equipment case.* For example, the signal grounds on printed circuit boards must not be connected to the chassis. On the other hand, if the designer determines that the metal chassis can be used as a signal reference for the low-frequency circuits without creating interference problems, and it is desirable to do so, the chassis then must be floated from the equipment case through the use of insulating spacers or standoffs. Care must be exercised in the mechanical layout of the equipment to insure that screws and fasteners do not compromise this isolation.

Controls, readout and indicating devices, fuses and surge protectors, monitoring jacks, and signal connectors must be installed so that they do not compromise this isolation. Both sides of the ac power line must be isolated from the low-frequency signal ground and from the equipment case. Only transformer-type power supplies should be used; the commercial ac/dc practice should *never* be used. (The metal portions of equipments exposed to human contact must be grounded with the green safety wire in accordance with the National Electrical Code).

---

* Common battery systems typically are designed with the signal ground connected to chassis or cabinet ground which is also connected to one of the dc supply buses. It is recommended that such systems be isolated from the structure and from the racks and cabinets of other low frequency equipment and systems. All interfaces between common battery systems and other equipment and systems must be balanced. Shield grounding must be controlled to ensure that the desired isolation is maintained.

7.7 High-Frequency Circuit Grounding[7]

Because of the impedance properties of practical sized conductors, meaningful single-point grounding cannot realistically be achieved at high frequencies. The inherent (ac) resistance, inductance, and capacitance of conductors produce impedance effects that become integral elements of circuit components, and circuit performance becomes difficult to predict. To reduce inductance effects, as much metallic surface as possible is typically sought for circuit common. All components are referenced to this common. Where capacitive-related feedback exists in excessive amounts, frequently it can be controlled through the insertion of shields between sensitive elements. A reliable design procedure for HF through microwave circuits is to segment separate stages into self-contained metal compartments to minimize stray capacitive coupling.

Multiple point grounding is more practical than single-point grounding in high frequency equipment. The various signal pairs internal to the equipment are referenced as required to a metallic common, or ground plane, with minimum length conductors. The equipment chassis is grounded through the case or cabinet to the facility ground system, i.e., the structure. If the high-frequency circuits are sensitive to structural ground currents (as they might be in the situation where multioctave bandwidths are involved), it may be necessary to construct the circuitry on a subchassis and isolate the subchassis from the cabinet except for one connection. Where two equipment elements of this type must be interconnected with unbalanced cables, either one end only must be grounded while the other is left floating or appropriate isolation in the signal path must be provided to break the ground loop.

For high-frequency signals, the interfacing lines between equipment elements will normally be unbalanced, constant impedance, transmission lines such as coaxial cables. The current return conductor, e.g., the shield in the case of a standard coaxial cable, should be grounded to the equipment enclosure at both ends of the cable and at intermediate points along the cable run. (This multiplepoint grounding of the shield maintains the RF shielding effectiveness of the cables and simplifies equipment design).

Cable connectors must provide a low-impedance path between the cable shield and the equipment case on which the connector is mounted. Bond the shield completely around the periphery of the cable to the connector shell with a tight compression or soldered bond (soldered connections are preferred over clamps). High-frequency shield terminations must maintain the RF tightness of the interconnected system (see Fig. 7.18).

If direct grounding of high-frequency circuits and subassemblies to the chassis is not desirable, a shielded, constant impedance cable

Sec. 7.7   High-Frequency Circuit Grounding

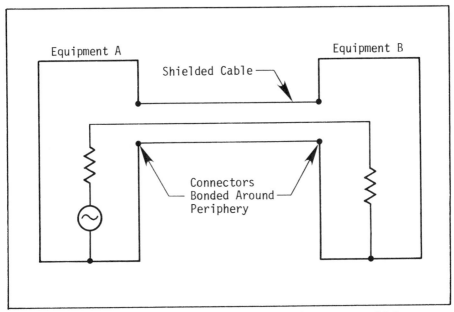

Figure 7.18 - Establishment of Shield Continuity Between High Frequency Equipments.

must be used that does not use the outer shield as a single return. For example, in low level, wideband (particularly video) systems, noise voltages from currents induced in cable shields by incident RF fields, i.e., the antenna effect, or arising from stray power currents and flowing through the cable shield can be troublesome. A way to combat the RF pickup problem is to enclose the shield carrying the signal return current inside of another shield or use a balanced type of transmission line. To accomplish the first of these alternatives, either a triaxial type cable can be used or the coaxial cable can be enclosed in metallic conduit. The inner shield of the triaxial cable, or the shield of the conduit-protected coaxial cable, should be terminated at the signal ground on the inside of the equipment. The outer shield of the triaxial cable and the conduit should be peripherally bonded to the case or cabinet of the terminating equipments. If the interference is the result of stray power currents, the current path through the shield must be interruped or a twin-axial type of cable must be used. To interrupt the path for stray power currents, the system's signal reference must be connected to the structure at only one end. Thus, either the source or load end signal reference must be isolated from the structure and the ac ground. Isolation can be achieved either by floating the equipment or its internal circuitry. (Generally, either process is very difficult to implement and maintain. It is preferable to resort to a balanced interface or locate the interference source and reduce the magnitude of the stray current).

Sec. 7.7 Hybrid Circuits

Proper connectors that do not compromise the dc isolation must be employed. For example, triaxial cable interfaces require the use of triaxial connectors to maintain isolation between the two shields.

7.8 Hybrid Circuits

Some types of equipment necessarily will contain both low and high frequency signal circuits in the same enclosure because of specific design or operational requirements. (For example, a typical VHF or UHF receiver will require both a high-frequency input from the antenna and a low-frequency output to audio or IF amplifiers as illustrated in Fig. 7.19). The high frequency interfaces to all transitional type equipment should be constant impedance shielded lines to the chassis or cabinet with the shield grounded around its periphery. The low frequency interfaces should be shielded, balanced, twisted-pair lines as illustrated with the shield grounded only at one end. *However, in equipment where both low and high-frequency circuits must share a common signal ground, because of design or construction requirements, both signal circuits should be grounded as in high frequency equipment.*

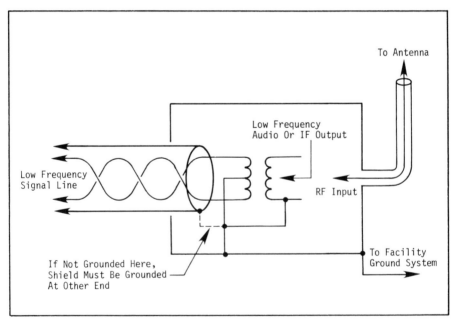

Figure 7.19 - Grounding Practices in Equipments Containing Both High Frequency and Low Frequency Circuits.

7.9 Shield Grounding

All individual shields of low-frequency signal lines within a cable bundle must be insulated from each other to minimize cross coupling. Further, these individual shields must be isolated from the overall bundle shield, equipment chassis and enclosures, junction boxes, conduit, cable trays, and all other elements of the facility ground system. When cables are long, extra attention must be directed toward maintaining the isolation of the individual shields at the ungrounded end and at all intermediate connectors throughout the cable run.

Shields of individual low frequency signal lines at terminating equipment may be carried into the case or cabinet on separate pins or they may be grounded together to be carried in (or out) in a common connector pin, depending upon the characteristics of the equipment involved. If the common pin arrangement is used, it must not compromise the single-point grounding principle. It is advisable to use one pin for low-level signal shields with a different pin used for high level signal lines. These individual shields should be terminated to the low frequency signal ground network. Where shield pigtails must be used, the pigtail between the shield breakout and the connector pin should be as short as practical.

Some of the individually shielded signal lines in multiconductor cables will be grounded at the end while other shields will be grounded at the other end. Careful attention must be given to the installation of such cables to prevent grounding of shields at both ends. Multiconductor cables which contain unshielded or individual shielded wires, or both, frequently have an overall shield provided for both physical protection and to provide supplemental electromagnetic shielding. Such overall shields should be grounded at each and of the cable run to provide a continuous RF shield with no breaks.

On long cable runs where the cable is routed through one or more intermediate connectors, the overall shield should be grounded to the frame or case of junction boxes, patch panels, and distribution boxes along the cable run.

For maximum shielding effectiveness, the overall shield should be effectively bonded with a low impedance connection to the equipment case, enclosure wall, or other penetrated (metal) shield as shown in Fig. 7.20. The best way to bond the overall shield to a connector is to run the shield wall inside of the connector shell and provide clean metal-to-metal circumferential contact between the shield and the shell. If the connector is not involved, the shortest practical lengths of connecting strap or jumper should be used. Where the over all shield terminates on a terminal strip, it may be grounded as shown in Fig. 7.21.

Sec. 7.9 Shield Grounding

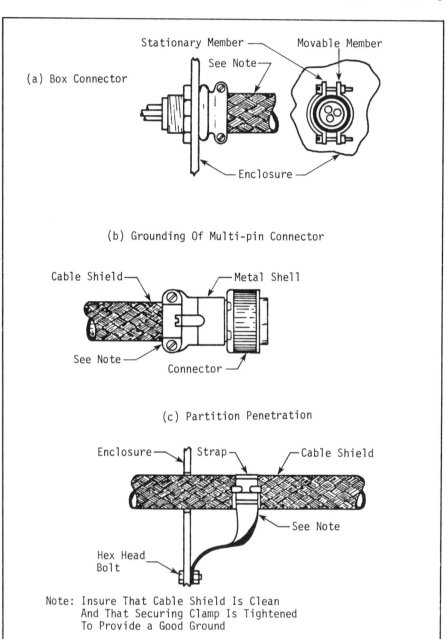

Figure 7.20 - Grounding of Overall Cable Shields to Connectors and Penetrated Walls.

Sec. 7.9 Shield Grounding

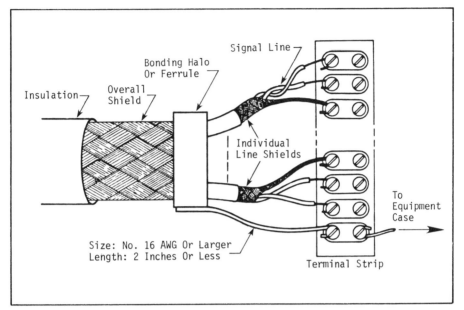

Figure 7.21 - Grounding of Overall Cable Shields to Terminal Strips.

7.10 References

1. Brown, Harry, *Don't Leave System Grounding to Chance*, EDN/EEE, January 15, 1972, pp. 22-27.

2. Walker, William T., *Test Equipment Grounding*, FAA/Georgia Institute of Technology Workshop on Grounding, Bonding and Shielding, Report No. FAA-RD-76-104, May 1976.

3. Nalle, D., *Elimination of Noise in Low-Level Circuits*, ISA Journal, August 1965, pp. 59-68.

4. Schneider, D.B., *Designing for Low Level Inputs*, Electronic Industries, January 1961, pp. 81-85.

5. *Instrumentation Grounding and Noise Minimization Handbook*, AFRPL-TR-65-1, Consolidated Systems Corporation, Pomona, CA, January 1965, AD 612-027.

6. Morrison, Ralph, *Grounding and Shielding Techniques in Instrumentation*, John Wiley and Sons, New York, NY (1967).

7. Denny, H.W., and Woddy, J.A., *Grounding, Bonding and Shielding Practices and Procedures for Electronic Equipments and Facilities*, Volume II - Procedures for Facilities and Equipments, Report No. FAA-RD-75-315, II, Engineering Experiment Station, Georgia Institute of Technology, Atlanta, GA, December 1975, AD A022 608.

CHAPTER 8

Bonding

Bonding is the process by which a low impedance path for the flow of an electric current is established between elements of the grounding network, or between joints in a shield. In an electronic system, numerous interconnections between conductors must be made in order to provide electric power, minimize electric shock hazards, provide lightning protection, establish references for electronic signals, etc. Ideally, each of these interconnections should be made so that the mechanical and electrical properties of the path are determined by the connected members and not by the joint. The joint must maintain its properties over an extended period of time. This chapter reviews various direct bonding techniques, discusses the frequency dependent properties of indirect bonds, presents the techniques for preventing bond corrosion, and provides a generalized set of bonding practices appropriate for electronic equipment.

8.1 Effects of Poor Bonds

Poor bonds lead to a variety of hazardous and interference-producing situations. For example, loose connections in ac power lines may cause heat to be generated in the joint and damage the insulation of the wires. Loose or high impedance joints in signal lines are particularly annoying because of intermittent signal behavior such as decreases in signal amplitude, increases in noise level, or both. Degradations in system performance from high noise levels are frequently traceable to poorly bonded joints in circuit returns and signal referencing networks.

Bonding is also important to the performance of other interference control measures. For example, adequate bonding of connector shells to equipment enclosures is essential to the maintenance of the integrity of cable shields and to the retention of the low-loss transmission properties of the cables. The careful bonding of seams and joints in enclosures and covers is essential to the achievement of a high degree of shielding effectiveness. Interference reduction

Sec. 8.1 Effects of Poor Bonds

components and devices also must be well bonded for optimum performance. For example, consider a typical power-line filter like that shown in Fig. 8.1. If the return side of the filter (usually the housing) is inadequately bonded to the ground reference plane (typically the equipment case or rack), the bond impedance Z_B may be high enough to impair the filter's performance. If Z_B is high relative to reactance X_C, interference currents will follow Path 2 to the load and the effectiveness of the filter will be compromised. Poor bonds in the presence of high-level RF fields, such as those in the immediate vicinity of high powered transmitters, can produce a particularly troublesome type of interference. Poorly bonded joints have been shown to generate cross modulation and intermodulation products when irradiated by two or more high level signals.[1]

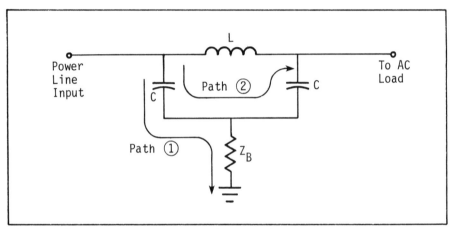

Figure 8.1 - Effects of Poor Bonding on the Performance of a Power Line Filter.

8.2 Bond Resistance

A primary requirement for effective bonding is that a low resistance path be established between two joined objects. A bonding resistance of 1 milliohm indicates a high quality junction.[2] Experience shows that 1 milliohm can be achieved if surfaces are properly cleaned and adequate pressure is maintained between the mating surfaces. There is little need to strive for a junction resistance that is appreciably less than the intrinsic resistance of the conductors being joined.

A similarly low value of resistance between widely separated points on a ground reference plane or network insures that all junctions are well made and that adequate quantities of conductor are provided throughout the plane or network. In this way, resistive voltage drops are minimized which enhance noise control.

It should be recognized that a low dc bond resistance is not a reliable indicator of the performance of the bond at high frequencies. Inherent conductor inductance and stray capacitance, along with associated standing wave effects and path resonances, will determine the impedance of the bond. Thus, in RF bonds, these factors must be considered along with the dc resistance.

8.3 Direct Bonds

Direct bonding is where specific portions of the surface areas of the members are placed in direct contact. Electrical continuity is obtained by establishing a fused metal bridge across the junction by welding, brazing, or soldering or by maintaining a high pressure contact between the mating surfaces with bolts, rivets, or clamps. Properly constructed direct bonds exhibit a low dc resistance and provide an RF impedance as low as the configuration of the bond members will permit. Direct bonding is always preferred; it can be used only when the two members can be connected together without an intervening conductor and can remain so without relative movement.

Direct bonds may be either permanent or semipermanent in nature. Permanent bonds may be defined as those intended to remain in place for the expected life of the installation and not required to be disassemblied for inspection, maintenance, or system modifications. Joints which are inaccessible by virtue of their location should be permanently bonded and appropriate steps taken to protect the bond against deterioration.

Many bonded junctions must retain the capability of being disconnected without destroying or significantly altering the bonded members. Junctions which should not be permanently bonded include those which may be broken for system modifications, for network noise measurements, for resistance measurements, and for other related reasons. In addition, many joints cannot be permanently bonded for reasons of costs. All such connections not permanently joined are defined as semipermanent bonds. Semipermanent bonds include those which utilize bolts, screws, rivets, clamps, and other auxiliary devices for fasteners.

8.3.1 Welding

In terms of electrical performance, welding is the ideal method of bonding. The intense heat involved is sufficient to boil away contaminating films and foreign substances. A continuous metallic bridge is formed across the joint; the conductivity of this bridge approximates that of the primary members. The net resistance of the bond is essentially zero because the bridge is very short relative to the length of the bond members. The mechanical strength of the bond is high; the strength of a welded bond can approach or exceed the strength of the bond members themselves. Since no moisture or contaminants can penetrate the weld, bond corrosion is minimized.

8.3.2 Brazing

Brazing (to include silver soldering) is another metal flow process for permanent bonding. As with welds, the resistance of the brazed joint is essentially zero. Since brazing frequently involves the use of a metal different from the primary bond members, precaution must be taken to protect the bond from deterioration through corrosion.

8.3.3 Soft Solder

Soft soldering is attractive because of the ease with which it can be applied. Properly applied to compatible materials, the bond provided by solder is nearly as low in resistance as one formed by welding or brazing. Because of its low melting point, however, soft solder should not be used as the primary bonding material where high currents may be present, as in power fault or lightning discharge paths.

8.3.4 Bolts

In many applications, permanent bonds are not desired. The most common semipermanent bond is the bolted connection (or one held in place with machine screws, lag bolts, or other threaded fasteners) because this type of bond provides the flexibility and accessibility that is frequently required. The bolt (or screw) should serve only as a fastener to provide the necessary force to maintain the 1200-1500 psi pressure required[3] between the contact surfaces for satisfactory bonding. Except for the fact that metals are generally necessary to provide tensile strength, the fastener does not have to be conductive.

8.3.5 Conductive Adhesive

Conductive adhesive is a silver-filled, two-component, thermosetting epoxy resin which, when cured, produces an electrically conductive material. It can be used between mating surfaces to provide low resistance bonds. It offers the advantage of providing a direct bond without the application of heat. When used in conjunction with bolts, conductive adhesive provides an effective metal-like bridge with high mechanical strength. It should be used with care, however, for there are indications that its properties may deteriorate with time.

8.4 Indirect Bonds

Operational requirements or equipment locations often preclude direct bonding. When physical separation is necessary between the elements of an equipment complex or between the complex and its reference, auxiliary conductors must be incorporated as bonding straps or jumpers. Such straps are commonly used for bonding of shock mounted equipment to the structural ground reference. They are also used for bypassing structural elements, such as the hinges on distribution box covers or on equipment covers, to eliminate the wideband noise generated by those elements when illuminated by intense radiated fields or when carrying high level currents. Bond straps or cables are also used to prevent static charge buildup and to connect metal objects to lightning down conductors to prevent flashover.

The geometrical configuration of the bonding conductor and the physical relationship between objects being bonded introduce reactive components into the impedance of the bond. The strap itself exhibits an inductance that is related to its dimensions. The inductive reactance of representative types of bonding conductors is plotted in Fig. 8.2 as a function of frequency. At relatively low frequencies, the reactance of the inductive component of the bond impedance becomes much larger than the resistance.[3,4] Thus, in the application of bonding straps, the inductive properties, as well as the resistance of the strap, must be considered.

The physical size of the bonding strap is important because of its effect on the RF impedance. As the length of the strap is increased, its impedance increases linearly for a given width; however, as the width increases, there is a nonlinear decrease in strap impedance. In Chap. 2, it was shown that the relative reactance of a strap decreases significantly as the length to width (ℓ/b) ratio decreases. Because of this reduction in reactance, bonding straps which are expected to provide a path for RF currents are frequently recommended to maintain a length-to-width ratio of 5 to 1 or less, with a ratio of 3 to 1 preferred.

In many applications, braided straps are preferred over solid straps because they offer greater flexibility for a given dc resistancy. Tests[4] confirm that there is no essential difference in the RF impedance properties of braided and solid straps of the same dimensions and made of the same materials. Because the strands are exposed, they are more susceptible to corrosion and fraying; thus, braided straps may be undesirable for use in some locations.

A certain amount of stray capacitance is inherently present between the bonding jumper and the objects being bonded, and between the bonded objects themselves. Figure 8.3 shows the equivalent circuit of an indirectly bonded system. The bonding strap parameters are represented by R_s, C_s, and L_s.

Sec. 8.4   Indirect Bonds

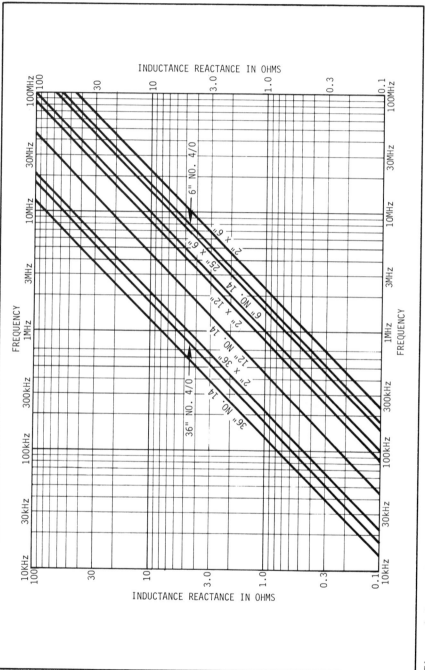

Figure 8.2 - Inductive Reactance of Wire and Strap Bond Jumpers.

Sec. 8.4   Indirect Bonds

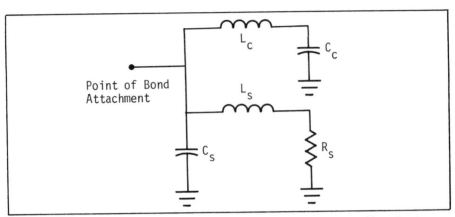

Figure 8.3 - True Equivalent Circuit of an Indirect Bonding System.

The inherent inductance of a bonded object, e.g., an equipment rack or cabinet, is represented by L_c and the capacitance between the bonded members, i.e., between the equipment and its reference plane, is represented by C_c. In most situations, L_s is much larger than L_c, C_c is much greater than C_s, and R_s can be ignored. Thus, the primary (i.e., the lowest) resonant frequency is given by:

$$f_r = \frac{1}{2\pi\sqrt{L_s C_s}} \qquad (8.1)$$

These resonances can occur at surprisingly low-frequencies--as low as 10 to 15 MHz[3] in typical configurations. In the vicinity of these resonances, bonding path impedances of several hundred ohms are common. Because of such high impedances, the strap is not effective. In fact, in these high impedance regions, the bonded system may act as an effective antenna system which increases the pickup of the same signals which bond straps are intended to reduce. In general, it has been shown that:

- at low frequencies, where the reactance of the strap is low, bonding straps will provide effective bonding,

- at frequencies where parallel resonances exist in the bonding network, straps may severely enhance the pickup of unwanted signals,

- above the parallel resonant frequency, bonding straps do not contibute to the pickup of radiated signals either positively or negatively.

In essence, bonding straps should be selected and used with care with special note taken to assure that unexpected interference conditions are not generated by the use of such straps.

8.5 Bond Corrosion

Corrosion is the deterioration of a substance (usually a metal) because of a reaction with its environment. Most enviroments are corrosive to some degree. Bonds exposed to these environments must be protected to prevent deterioration of the bonding surfaces to the point where the required low resistance connection is destroyed.

8.5.1 Chemical Basis of Corrosion

The requirements for corrosion to take place are (1) an anode and a cathode must be present to form an electrochemical cell and (2) a complete path for the flow of direct current must exist. These conditions occur readily in many environments. On the surface of a single piece of metal, anodic and cathodic regions are present because of impurities, grain boundaries and grain orientations, or localized stresses. These anodic and cathodic regions are in electrical contact through the body of metal. The presence of an electrolyte or conducting fluid completes the circuit and allows the current to flow from the anode to the cathode of the cell.

Anything that prevents the existence of either of these conditions will prevent corrosion. For example, strive to prevent the joining of dissimilar metals (e.g., aluminum/copper, aluminum/steel) in damp environments or thoroughly protect the completed joint with paint or other sealant.

When joints between dissimilar metals are unavoidable, the anodic member (see Table 8.1) of the pair should be the largest of the two. For a given current flow in a galvanic cell, the current density is greater for a small electrode than for a larger one. The greater the density of the current leaving an anode, the greater is the rate of corrosion. As an example, if a copper strap is bonded to a steel column, the rate of corrosion of the steel will be low because of the large anodic area. On the other hand, a steel strap or bolt fastener in contact with a copper plate will corrode rapidly because of the relatively small area of the anode of the cell.

Paint or metallic platings used for the purpose of excluding moisture or to provide a third metal compatible with both bond members should be applied with caution. When they are used, both members must be covered as illustrated in Fig. 8.4. Covering the anode alone must be avoided. If only the anode is covered, then at imperfections and breaks in the coating, corrosion will be severe

Sec. 8.5   Bond Corrosion

Table 8.1 - Galvanic Series of Common Metals and Alloys[5]

(ANODIC OR ACTIVE END)
Magnesium
Magnesium Alloys
Zinc
Galvanized Steel or Iron
1100 Aluminum
Cadmium
2024 Aluminum
Mid Steel or Wrought Iron
Cast Iron
Chromium Steel (active)
Ni-Resist (high-Ni cast iron)
18-8 Stainless Steel (active)
18-8 Mo Stainless Steel (active)
Lead-tin Solders
Lead
Tin
Nickel (active)
Inconel (active)
Hastelloy B
Manganese Bronze
Brasses
Aluminum Bronze
Copper
Silicon Bronze
Monel
Silver Solder
Nickel
Inconel
Chromium Steel
18-8 Stainless Steel
18-8 Mo Stainless Steel
Hastelloy C
Chlorimet 3
Silver
Titanium
Graphite
Gold
Platinum
(CATHODIC OR MOST NOBLE END)

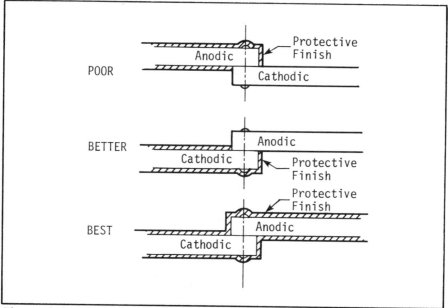

Figure 8.4 - Techniques for Protecting Bonds Between Dissimilar Metals.

because of the relatively small anode area. All such coatings must be maintained in good condition.

8.5.2 Bond Protection Code

For bonds of high reliability, corrosion must be prevented by (1) avoiding the pairing of dissimilar metals and (2) preventing the entrance of moisture or other electrolytes into the bond area. Metals to be in direct contact should fall as close together in the galvanic series as possible. Compatible groupings of the common metals are given in Table 8.2. The corrosive action between metals of different groups will be the greatest when the joint is openly exposed to salt spray, rain, or other liquids.

Table 8.2 - Compatible Groups of Common Metals

Group	Metals
I	Magnesium
II	Aluminum, aluminum alloys, zinc, cadmium
III	Carbon steel, iron, lead, tin, lead-tin solder
IV	Nickel, chromium, stainless steel
V	Copper, silver, gold, platinum, titanium

Sec. 8.5   Bond Corrosion

The relative degrees of exposure may be defined as follows:[6]

  Exposed: Open, unprotected exposure to weather.

  Sheltered: Limited protection from direct action of
                weather. Locations in louvered housings,
                sheds and vehicles offer sheltered exposure

  Housed: Located in weatherproof buildings.

When bonds under these different exposure conditions must be made
between different groups, they should be protected as indicated by
Table 8.3. *Condition A* in Table 8.3 means that the couple (joint)
must have a protective finish applied after metal-to-metal contact
has been established so that no liquid film can bridge the two elements of the couple. *Condition B* means that the two metals may be
joined with bare metal exposed at junction surfaces. The remainder
of the bond must be given an appropriate protective finish. *Condition
C* indicates that the combination cannot be used except under very unusual circumstances where short life expectancy can be tolerated or
when the equipment is normally stored and exposed for only short
intervals. Protective coatings are mandatory under these circumstances.

Sec. 8.5   Bond Corrosion

Table 8.3 - Bond Protection Requirements

Condition of Exposure	Anode				Cathode
	I	II	III	IV	
Exposed	A	A			
Sheltered	A	A			II
Housed	A	A			
Exposed	C	A	B		
Sheltered	A	B	B		III
Housed	A	B	B		
Exposed	C	A	B	B	
Sheltered	A	A	B	B	IV
Housed	A	B	B		
Exposed	C	C	C	A	
Sheltered	A	A	A	B	V
Housed	A	B	B	B	

A - Protective finish applied after metal-to-metal contact so any liquid film cannot bridge the two elements of the couple.

B - Two metals may be joined with bare metal exposed at junction surfaces. Remainder of bond must be given protective finish.

C - The combination cannot be used except under rare circumstances where short life expectance can be tolerated. Protective coatings are mandatory.

8.5.3 Jumper Fasteners

Acceptable fastener materials for bonding aluminum and copper jumpers to other metals are indicated in Table 8.4. The arrangement of the metals is in the order of decreasing galvanic activity. The screws, nuts and washers to be used in making the connection as indicated are:

Type I - Cadmium or zinc plated steel, or aluminum,

Type II - Passivated stainless steel.

Where either type of securing hardware is indicated, Type II is preferred from the corrosion standpoint.

Table 8.4 - Metal Connections for Aluminum and Copper Jumpers

Metal Structure (Outer Finish Metal)	Connection For Aluminum Jumper	Screw Type	Connection For Tinned Copper Jumper	Screw Type
Magnesium and Magnesium Alloys	Direct or Magnesium Washer	Type I	Aluminum or Magnesium Washer	Type I
Zinc, Cadmium, Aluminum and Aluminum Alloys	Direct	Type I	Aluminum Washer	Type I
Steel (except stainless steel)	Direct	Type I	Direct	Type I
Tin, Lead, and Tin-Lead Solders	Direct	Type I	Direct	Type I or II
Copper and Copper Alloys	Tinned or Cadmium Plated Washer	Type I or II	Direct	Type I or II
Nickel and Nickel Alloys	Tinned or Cadmium Plated Washer	Type I or II	Direct	Type I or II
Stainless Steel	Tinned or Cadmium Plated Washer	Type I or II	Direct	Type I or II
Silver, Gold and Precious Metals	Tinned or Cadmium Plated Washer	Type I or II	Direct	Type I or II

8.6 Workmanship

Whichever bonding method is determined to be the best for a given situation, the mating surfaces must be cleaned of all foreign material and substances which would preclude the establishment of a low resistance connection. Next, the bond members must be carefully joined employing techniques appropriate to the specific method of bonding. Finally, the joint must be finished with a protective coating to ensure continued integrity of the bond. The quality of the junction depends upon the thoroughness and care with which these three steps are performed.

Personnel making bonds must be carefully trained in the techniques and procedures required. Where bonds are to be welded, for example, work should be performed only by qualified welders. No additional training should be necessary because standard welding techniques appropriate for construction purposes are generally sufficient for establishing electrical bonds. Qualified welders should also be used where brazed connections are to be made.

Pressure bonds utilizing bolts, screws, or clamps must be given special attention. Usual construction practices do not require the surface preparation and bolt tightening necessary for an effective and reliable electrical bond. Therefore, emphasis beyond what would be required for strictly mechanical strength is necessary. Bonds of this type must be checked rigorously to see that the mating surfaces are carefully cleaned, that the bond members are properly joined, and that the completed bond is adequately protected against corrosion.

Specific recommendations are:

- Utilize welding whenever possible for permanently-joined bonds. The welds must be adequate to support the mechanical load demands on the bonded members.

- Use brazing (or silver soldering) for permanently bonding copper and brass.

- Do not use soldered connections in fault protection grounding networks or for any connection in the lightning protection system.

- The mating surfaces of bolted and other compression type bonds require careful cleaning. The basic cleaning requirements are:

    - All nonconductive material must be removed. Such materials include paints and other organic finishes; anodizing films; oxide and sulfide films; and oil, grease and other petroleum products.

- All corrosive agents must be removed. Such agents include water, acids, strong alkalies, and any other materials which provide conductive electrolytic paths.

- All solid matter which would interfere with the establishment of a low resistance path across the bond interface or which forms a wedge or barrier to keep the bond area open to the entrance of corrosive materials or agents must be removed. Such solid materials include dust, dirt, sand, metal filings, and corrosion by-products.

• The proper order of assembly for bolted bonds is illustrated in Fig. 8.5. Position load distribution washers directly underneath the bolt head or under the nut next to the primary member. Lockwashers may be placed between the nut and any load distribution washers. Toothed lockwashers should not be placed between the primary bonded members.

• Once the mating surfaces have been cleaned of all nonconductive material, join the bond members together as soon as possible. If delays beyond two hours are necessary in corrosive environments, the cleaned surfaces must be protected with an appropriate coating which, of course, must be removed before completing the bond.

• Alligator clips and other spring loaded clamps are to be employed only as temporary bonds. Use them primarily to ensure that personnel are not inadvertently exposed to hazardous voltages when performing repair work on equipment or on facility wiring.

Sec. 8.6 　　　　　　　　　　　　　　　　　　　　　　　　　　　　Workmanship

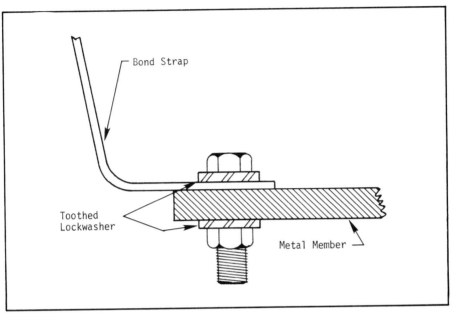

Figure 8.5 - Order of Assembly for Bolted Connection.

8.7 Equipment Bonding Practices

Bonding is an essential element of the equipment interference control effort. This section presents design and construction guidelines to aid in the implementation of effective bonding of equipment circuits, equipment enclosures, and cabling. These guidelines are not intended as step-by-step procedures for meeting EMC specifications. Rather they are aimed at focusing attention on those principles and techniques which lead to increased compatibility between circuits, assemblies and equipments:

- Welded seams should be used wherever possible because they are permanent, offer a low-impedance bond, and provide the highest degree of RF shielding.

- Spot welds may be used where RF tightness is not necessary. Spot welding is less desirable then continuous welding because of the tendency for buckling and the possibility of corrosion occuring between welds.

- Soldering should not be used where high mechanical strength is required. If mechanical strength is required, the solder should be supplemented with fasteners such as screws or bolts.

- Fasteners such as bolts, rivets, or screws should not be relied upon to provide the primary current path through a joint.

- Rivets should be used primarily to provide mechanical strength to soldered bonds.

- Sheet metal screws should be used only for the fastening of dust covers on equipment or for the attachment of covers to discourage unauthorized access by untrained personnel.

- The following precautions should be observed when employing bonding straps or jumpers:

  • Jumpers should be bonded directly to the basic structure rather than through an adjacent part.

  • Jumpers should not be installed two or more in series.

  • Jumpers should be as short as possible.

Sec. 8.7   Equipment Bonding Practices

- Jumpers should not be fastened with self-tapping screws.

- Jumpers should be installed so that vibration or motion will not affect the impedance of the bonding path.

• Where electrical continuity across shock mounts is necessary, bonding jumpers should be installed across each shock mount. Jumpers for this application should have a maximum thickness of 0.063 cm so that the damping efficiency of the mount is not impaired. In severe shock and vibration environents, solid straps may be corrugated or flexible wire braid may be used.

• Where RF shielding is required and welded joints can cannot be used, the bond surfaces must be machined smooth to establish a high degree of surface contact throughout the joint area. Fasteners must be positioned to maintain uniform pressure throughout the bond area.

• Chassis-mounted subassemblies should utilize the full mounting area for the bond as illustrated in Fig. 8.6 and 8.7. Separate jumpers should not be used for this purpose.

• Equipment attached to frames or racks by means of flange-mounted quick disconnect fasteners must be bonded about the entire flange periphery as shown

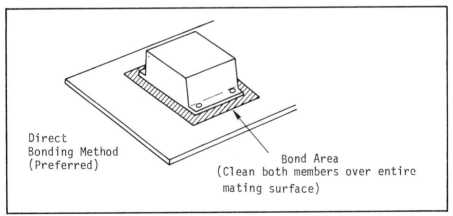

Figure 8.6 - Bonding of Subassemblies to Equipment Chassis.

Sec. 8.7   Equipment Bonding Practices

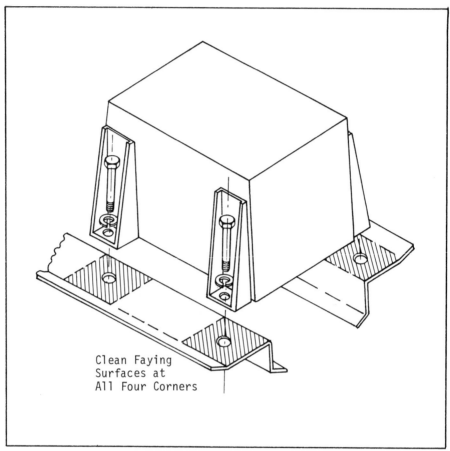

Figure 8.7 - Bonding of Equipment to Mounting Surface.

      in Fig. 8.8. Both the flange surface and the mating rack surface must be cleaned of paint and other foreign materials over the entire contact area.

- Rack mounted packages employing one or more dagger pins should be bonded as shown in Fig. 8.9.

- The recommended practices for effective bonding of equipment racks are shown in Fig. 8.10. Bonding between the equipment chassis and the rack is achieved through contact between the equipment front panel and the rack front brackets. These brackets are bonded to the horizontal slide which is in turn welded to the rack frame.

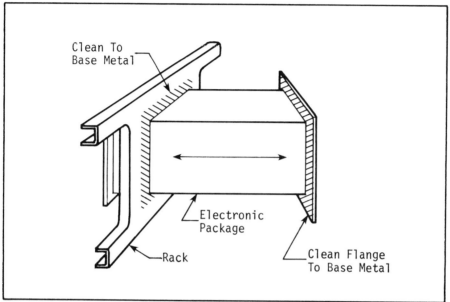

Figure 8.8 - Typical Method of Bonding Equipment Flanges to Frame or Rack.

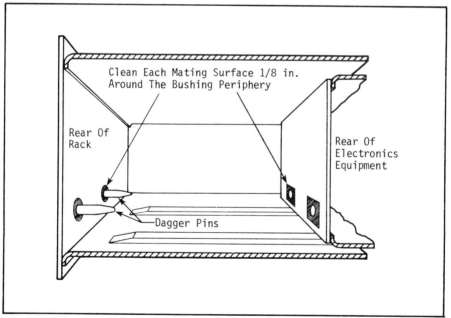

Figure 8.9 - Bonding of Rack-Mounted Equipment Employing Dagger Pins.

Sec. 8.7   Equipment Bonding Practices

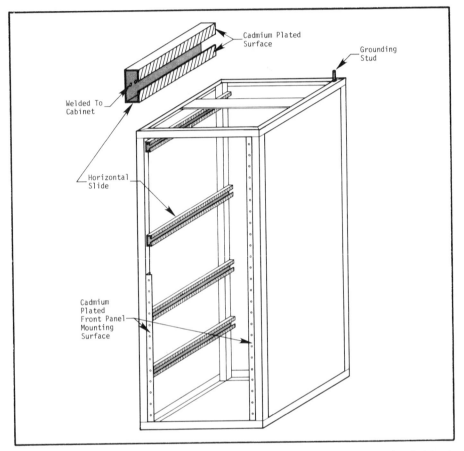

Figure 8.10 - Recommended Practices for Effective Bonding in Cabinets.

- Standard MS-type connectors and coaxial connectors must be bonded to their respective panels over the entire mating surface as illustrated in Fig. 8.11. Panel surfaces must be cleaned to the base metal for no less than 0.3 cm beyond the periphery of the mating connector.

- Individual shields on low frequency signal lines should be soldered to the appropriate pins of the connector. If solderless terminals must be used, a compressed fitting that provides maximum contact between the shield and the terminal sleeve should be used. Pigtails on individual low frequency signal line shields should be kept as short as possible.

Sec. 8.7   Equipment Bonding Practices

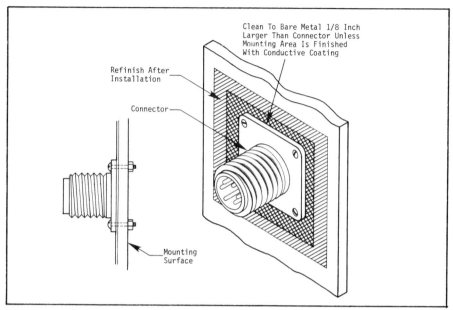

Figure 8.11 - Bonding of Connector to Mounting Surface.

- The overall cable shields on low frequency cables and the shields of high frequency signal lines must be bonded to the connector shell completely around the periphery of the shield with either compression or, preferably, soldered bonds.

- When an RF-tight joint is required at seams, access covers, removable partitions, and other shield discontinuities, conductive gaskets should be used. They may also be used to improve the bond between irregular or rough bonding surfaces. Gaskets should be sufficiently resilient to allow for frequency opening and closing of the joint and yet be stiff enough to penetrate any nonconductive films on surfaces.

- Gaskets should be firmly affixed to one of the bond members with screws, conductive cement, or any other means which does not interfere with their operation. The gaskets may be placed in a milled slot to prevent lateral movement.

- All bonds must be protected from corrosion and mechanical deterioration. Corrosion protection should be provided by insuring galvanic compatibility of metals and by sealing the bonded joint against moisture.

8.8 Summary of Bonding Principles

- Bonds must be designed into the grounding system. Specific attention should be directed to the interconnections not only in power lines and signal lines, but also between conductors of signal ground bus networks, between both cable and component or compartment shields and the ground reference plane, between structural members, and between elements of the lightning protection network. In the design and construction of a facility or an equipment, signal path, personnel safety, and lightning protection bonding requirements must be considered along with mechanical and operational needs.

- Bonding must achieve and maintain intimate contact between metal surfaces. The surfaces must be smooth and clean and free of nonconductive finishes. Fasteners must exert sufficient pressure to hold the surfaces in contact in the presence of the deforming stresses, shocks, and vibrations associated with the equipment and its environment.

- The effectiveness of the bond depends upon its construction, the frequency and magnitude of the currents flowing through it, and the environmental conditions to which it is subjected.

- Bonding jumpers are only a substitute for direct bonds. If the jumpers are kept as short as possible, have a low resistance and low ℓ/w ratio, and are not higher in the electrochemical series than the bonded members, they can be considered a reasonable substitute.

- Bonds are always best made by joining similar metals, If this is not possible, special attention must be paid to the control of bond corrosion through the choice of the materials to be bonded, the selection of supplementary components (such as washers) to assure that corrosion will affect replaceable elements only, and the use of protective finishes.

- Protection of the bond from moisture and other corrosion elements must be provided.

- Finally, throughout the lifetime of the equipment, system, or facility, the bonds must be inspected, tested, and maintained to assure that they continue to perform as required.

8.9 References

1. Henkel, R., and Mealey, D., *Electromagnetic Compatibility Operational Problems Aboard the Apollo Spacecraft Tracking Ship*, IEEE Electromagnetic Compatibility Symposium Record, IEEE 27080, July 1967, p. 70.

2. Denny, H.W., et al, *Grounding, Bonding, and Shielding Practices and Procedures for Electronic Equipments and Facilities*, Vol. I, Fundamental Considerations, Report No. FAA-RD-215, I, Contract No. DOT-FA72WA-2850, Engineering Experiment Station, Georgia Institute of Technology, Atlanta, Georgia, December 1975.

3. Denny, H.W. and Warren, W.B., *RF Bonding Impedance Study*, RADC-TR-67-106, Contract AF30(602)-3282, Engineering Experiment Station, Georgia Institute of Technology, Atlanta, Georgia, March 1967.

4. Evans, R.W., *Metal-to-Metal Bonding for Transfer of Radio Frequency Energy*, IN-R-ASTR-64-15, NASA Marchall Space Flight Center, Huntsville, Al., June 25, 1964.

5. Roessler, G.D., *Corrosion and the EMC/RFI Knitted Wire Mesh Gasket*, Frequency Technology, Vol. 7, No. 3, March 1969, pp. 15-24.

6. Taylor, R.E., *Radio Frequency Interference Handbook*, NASA-SP-3067, NSA, Washington, DC, 1971, N72-11153-156.

CHAPTER 9

Ground System Tests and Maintenance

A basic ingredient of engineering is the performance and interpretation of measurements. Measurements typically provide a great deal of insight into the operation of a device or system. Unfortunately, because of the wide frequency range over which a grounding network is capable of influencing system behavior and the physical extent (often many wavelengths long) of a network, effective measurements that provide results directly related to network performance are somewhat limited. One of the reasons that ground system design and analysis tend to be viewed as art rather than science is perhaps the unavailability of a comprehensive measurement process. For example, throughout this book it has been repeatedly emphasized that dc and low-frequency properties do not define the high-frequency properties of a grounding network. Yet, low-frequency measurements are the only easy and unambiguous ones to make--high-frequency measurement results include the effects of resonances, standing waves, and radiated pickup, all of which render interpretation difficult. However, indirect inferences can be made of high-frequency performance from low-frequency measurements with considerations of geometry and the other interference elements of coupling and shielding. The preceding chapters have established relationships between these factors and overall grounding network performance. Therefore, certain applicable measurements are set forth for use. In addition, some suggestions for interpreting the results are listed.

9.1 Test Procedures

Procedures for testing the performance of a grounding network at its frequencies of operation do not exist. Low frequency measurements can be made, however, and the results evaluated for comparative performance. The principal procedures available at the present time are bond resistance, ground system noise current, and differential noise voltage.

Sec. 9.1   Test Procedures

9.1.1 Bond Resistance

This test is intended to give a general indication of bond adequacy, based on the dc resistance of the bond.

- Equipment Requirements:

  - A dc resistance bridge capable of measuring to about 0.001 ohm or better. The bridge should be portable and not be position-sensitive. Connection of the test sample to the bridge terminals should be easily performed without cumbersome adapters of special tools. An instrument with separate potential (voltage) and current terminals is preferred to a two-terminal device.

  - A pair of heavy-duty spring clip leads for connection between the bridge and the bonded junction. Clip leads may be connected to braided straps and lugs to make connection to the bridge. Total resistance of external connectors and leads should not be greater than 0.001 ohm.

- Equipment Setup:

  - Using the heavy-duty spring clips and braids or low resistance wire, connect the leads to the bridge.

  - Place the bridge in operation according to the manufacturer's operation manual.

  - Zero the bridge, including leads, and connect the clip leads across the bonded junction as shown in Fig. 9.1. By placing the current leads away from the junction while placing the potential leads near the junction, the effects of the probe contact resistance are minimized. If the bond to be measured is internal to a metallic grid such that other current paths exist between the current probes in parallel to the one through the bond under test, the potential and current probes should be connected near the same point (one potential probe and one current probe on each side of the bond). Otherwise, a gross error may result. When multiple parallel paths exist, this procedure may not adequately indicate the true condition of the specific junction of interest but it will indicate the total resistance between both sides of the junction.

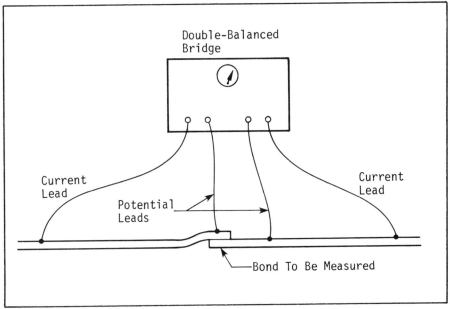

Figure 9.1 - Bond Resistance Measurement.

- Test Procedure:

    - Adjust the bridge balance until a null is obtained.

    - Record the indicated resistance.

9.1.2 Ground System Noise Current

This procedure measures the stray currents on safety grounds, signal grounds, and cable shields which are frequent causes of common mode noise interference within a facility.[1]

- Equipment Required:

    - Oscilloscope,

    - Oscilloscope current probe,

    - Oscilloscope camera,

- Equipment Setup and Test Procedure:

    - Connect the current probe, current probe amplifier, and oscilloscope as shown in Fig. 9.2.

Sec. 9.1

Test Procedures

- Observe the oscilloscope-displayed ambient level at each test point.

- Photograph the ambient level at each test point.

- Set the oscilloscope to trigger at a level slightly above the ambient.

- Set the oscilloscope for single sweep operation and open the camera shutter.

- Let the camera shutter remain open for 5 minutes or until the oscilloscope is triggered, whichever occurs first. (Longer sampling periods may be used if desired).

- Record pertinent information on the test photograph.

- If a spectrum analyzer plug-in is available for the oscilloscope, perform the current measurements in the frequency domain as well as the time domain. (Frequency domain measurements can be of great assistance in identifying the source of interference currents).

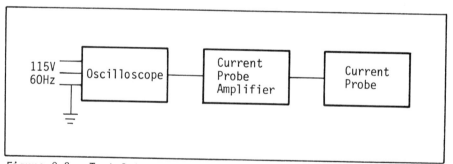

Figure 9.2 - Test Setup for Stray Current Measurements.

9.1.3 Differential Noise Voltage

- Equipment Required:

    - Oscilloscope with time and frequency domain plug-ins.

Sec. 9.1   Test Procedures

- Oscilloscope camera.

- Isolation transformer.

- Required lengths of shielded cable.

• Equipment Setup and Test Procedure:

- Set up the equipment as shown in Fig. 9.3. Note that the signal probe and the *ground* reference probe are connected to each of the two points between which the voltage differential is desired.

- After an adequate warm-up time, photograph the ambient noise level in both time and frequency domains. Figure 9.4 shows two examples of time domain results provided by this technique.

- If transient data are required, proceed as indicated in steps 4 through 7 in Sec. 9.1.2.

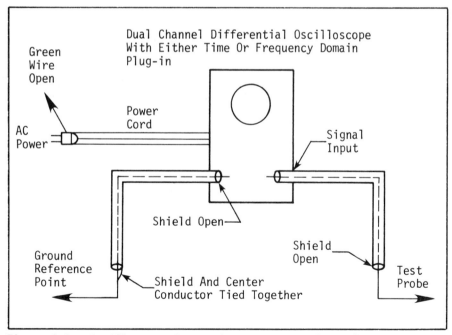

Figure 9.3 - Oscilloscope Connections for Measuring Voltage Levels on Ground Systems.

Vertical: 20mV/div.
Horizontal: 2ms/div.

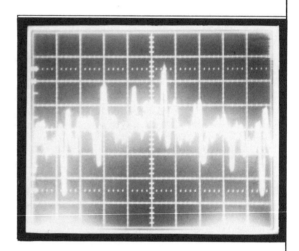

(a) Periodic Ground Network Noise.

Vertical: 10mV/div.
Horizontal: 2ms/div.

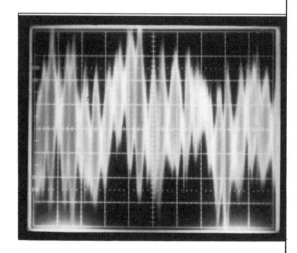

(b) RF Ground Network Noise.

Figure 9.4 - Typical Results Provided by Differential Noise Voltage Test.

9.2 Maintenance

Ground-system maintenance efforts can run the gamut from an occasional visual inspection to rigidly controlled periodic tests and measurements. The actual degree of maintenance complexity required for satisfactory performance is a function of many variables. Predominant among these are the ground system utilization, its location with regard to weather or corrosive environments, the materials from which it is constructed, etc.

9.2.1 Inspection

Ground system maintenance should include a periodic visual inspection of all possible interconnections and bonded joints as a minumum effort. During these inspections, indications of corrosion at the interface between mated metal surfaces and broken or disconnected ground conductors should be noted and corrected. Joints bonded together by bolts or clamps should specifically be inspected for corrosion. Emphasis should be directed to inspections for broken or disconnected ground conductors in areas of new construction and where flexible metal straps are used to bond movable devices to the ground system.

9.2.2 Bus System Requirements

In addition to visual inspections, it is desirable to conduct minimal tests to provide a qualitative indication of ground system adequacy.

9.2.2.1 Bonds

- Concurrent with or following the visual inspection of the bonds, perform bond resistance measurements. Select five to ten bonds that visually appear tight, well made, and corrosion free and measure their resistances. The sampling should include structural bonds, equipment-to-structure bonds, connections between safety ground wires, conduit-to-conduit or conduit-to-cabinet joints, bonds in lightning down conductors (to include structural columns if used for lightning discharge paths), and others as appropriate. Also measure all bonds exhititing visible defects.

- For every bond exhibiting a resistance greater than 1 milliohm, check for losseness; if the connections are loose, tighten the fastener. Measure the resistance again after tightening. If the resistance is still greater than 1 milliohm and the joint can be readily disassembled, disassemble the joint and check for corrosion, debris, paint, or other nonconductive materials. Remove the material, reassemble the bond, and remeasure the resistance.

- The grounded conductor, i.e., the safety, or green wire, is to have the grounding connection made only at the power service disconnecting means. Facilities which can be temporarily removed from service should be de-energized and the main power switch locked or otherwise secured open. With electrical power removed, disconnect the neutral from ground at the service disconnecting means and check for continuity between the neutral and the grounding conductor (see Fig. 9.5). A low resistance reading (< 10 ohms) indicates that the neutral is connected to ground somewhere other than at the service disconnect. This ground connection must be located and removed.

9.2.2.2 Network Resistance

Where low-frequency ground networks exist in a facility, measure the resistance between those points on the network where equipment interconnections are made. On Fig. 9.6, typical examples of this measurement are between equipment boxes D and E, between E and B, and between G and K. This resistance should not exceed 20 milliohms. Further, measure the resistance between the ground terminals of equipment that are also interconnected with signal cables and control lines. On systems employing equipment chassis or cabinet as signal ground (e.g., most high frequency and RF systems), measure the cabinet-to-cabinet (or chassis-to-chassis) resistance, particularly on those interconnected with signal cables (see connections B-F and F-H). Also measure the cabinet-to-structure resistance (see connections B-C and K-J). These two resistances should be less than 5 milliohms.

9.2.2.3 Stray Current Levels

Using a clamp-on ammeter, check the ac load currents on the conductors of three phase supply lines. Note particularly any differences in line currents greater than 10 percent.

Using a clamp-on ammeter measure the stray current levels in the safety ground network at selected points throughout the facility. Choose a sufficient number of points to give an indication of the relative stray current level in the facility. The current levels should be less than 0.1 ampere.[2]

Sec. 9.2   Maintenance

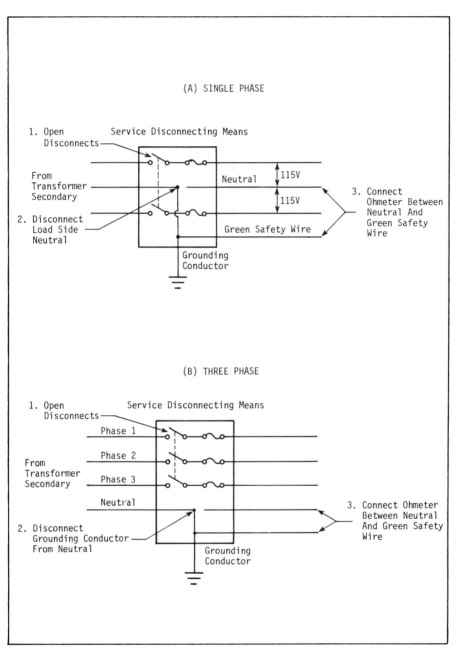

Figure 9.5 - Method for Determining the Existence of Improper Neutral Ground Connections.

Sec. 9.2 Maintenance

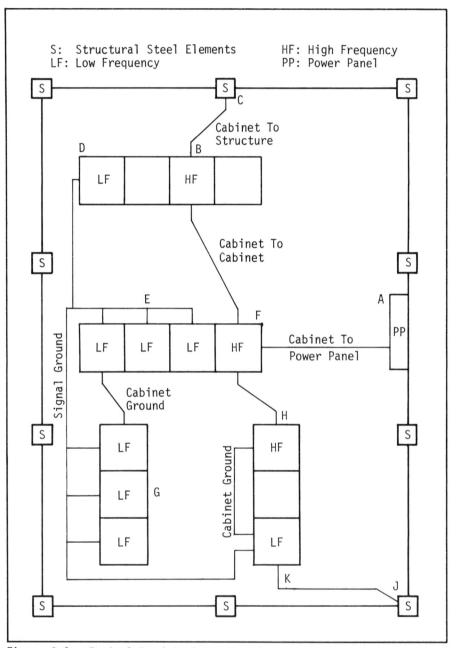

Figure 9.6 - Typical Bond Resistance and Stray Current Measurement Locations in an Electronic Facility.

Sec. 9.2   Maintenance

Using the clamp-on ammeter, probe signal ground wires, cable shields, or other conductors likely to be carrying stray power currents. Note particularly the current levels in the grounds of low frequency equipment and in the shields of cables carrying video, data, or other types of signals with operating frequencies in the power frequency range. (Typical locations in the illustration of Fig. 9.6 where stray current measurements should be made are on the connection B-C between the rack of high frequency equipment and the structure, on connection A-F between the equipment bay and the power panel, on the cabinet ground connection between low frequency equipment E and G, and on connection J-K).

9.2.2.4 Network Noise Level

In addition to visual inspections and dc resistance measurements, the signal levels being conducted through the ground system can also provide insight regarding ground system status.[3] Such signal levels are therefore an indication of the need, or lack of need, for ground system maintenance.

Using the test procedure of Sec. 9.1.2, measure the stray current levels on a selected number of shields surrounding sensitive signal cables, on conduit, and on equipment ground cables. Document the test details (i.e., vertical sensitivity and sweep rate) on the photographs made of the oscilloscope displays.

Using the test procedure described in Sec. 9.2.3, perform differential noise measurements between interfaced equipment; between the low frequency (or other) signal ground network and structural ground; between signal grounds in equipment and the point of connection to the earth electrode system; between widely separated points on ground networks; and between any other two points where common-mode voltages are causing system or equipment noise problems. (Oscilloscope displays should be retained in order to provide a relative indication of the ground system status as a function of time or system activity level. Signal levels in both time and frequency domains at ground points of interest during periods of minimum activity should be recorded. Then during periods of heavy activity, photographs of these same ground points are again made. The two photographs for each ground point should be retaken at regular intervals and immediately following major structural or equipment modifications or additions. This historical record will then permit short term compromises and long term deteriorations in the grounding system to be detected).

9.2.2.5 Star Ground Compromises

A measurement technique useful for identifying undesired short circuits between various elements of the single-point ground system is illustrated in Fig. 9.7. The technique, designated the current induction method[4], utilizes a convenient length of wire, a clamp-on ammeter, and a toroid coil through which 60 Hz current is passed. To determine if isolation is maintained between the signal ground and the facility ground, the auxiliary wire is threaded through the coil and is connected between the signal ground point and the facility ground. The normal connection between the signal ground and facility ground is then broken, as shown in Fig. 9.7a. A voltage is induced in the auxiliary wire and currents I_1, I_2, and I_3 are measured; if either of them is nonzero, the single-point ground system is compromised. The auxiliary wire is then connected between each chassis and the facility ground to determine where the compromise is, as indicated in Fig. 9.7b. Using the toroid coil, the current induced into each chassis is measured separately. The relative magnitude of these currents indicate the particular chassis in which the short circuit exists. For example, if the short circuit current induced in Chassis No. 3 is significantly larger than the currents induced in the other chassis, the compromise is at Chassis No. 3. (Removal of the undesired short circuit in Chassis No. 3 should be followed by a complete recheck of the entire system).

9.2.3 Corrective Action

The following guidelines may be used to help evaluate results of measurements and to help define the corrective actions which should be taken. This set of guidelines is not to be considered all inclusive. Specific situations can be expected to arise that will not be adequately covered by the guidelines. These situations must be recognized and dealt with on an individual basis.

- Review electrical wiring diagrams and the electrical equipment distribution within the facility to determine possible direct or indirect coupling paths between noisy equipment and susceptible electronic apparatus. Apply corrective measures such as:

  - relocate equipment,

  - redistribute the electrical load so that potentially interfering equipment units are served by separate feeders,

  - install electrical feeders in steel conduit or raceway to reduce magnetic fields,

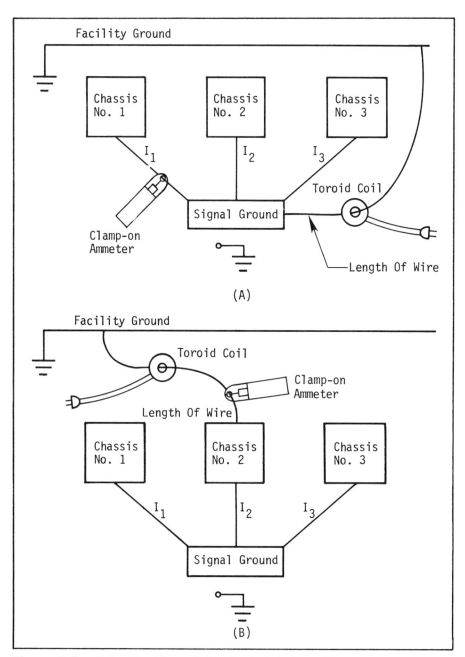

Figure 9.7 - Determination of Single-Point Ground System Compromises with the Use of Induced Currents.

Sec. 9.2   Maintenance

- relocate signal lines to sensitive equipment at the maximum possible distance from power conductors feeding noisy equipment.

● Correlate, if possible, any evidence of equipment malfunctions due to electrical noise on signal or control cables with the measured values of stray currents or voltages on grounding conductors and on cable shields. If such correlation exists, determine the probable cause of such noise voltages using the principles set forth earlier. Then apply the techniques described to reduce the noise to a level acceptable to the equipment in the facility.

● If operational experience as indicated by maintenance logs or outage reports and operator comments reveal problems with system noise and interference attributable to grounding deficiencies, choose the most appropriate noise minimization procedures and implement them.

● If more than 25 percent of the bonds measured exhibit a resistance greater than 1 milliohm, all bonds should be inspected carefully, and the resistance measured. Each one found to be deficient should be redone.

● Record all changes made during the maintenance process.

9.3 References

1. Zych, J.F., *Development of Simple Instrumentation to Measure and Assess Electronic Ground Systems*, FAA/GIT Grounding Workshop and Lightning Protection Seminar, FAA-RD-106, Atlanta, GA, May 1975, AD A013 618, pp. 45-62.

2. Denny, H.W., et al, *Grounding, Bonding and Shielding Practices and Procedures for Electronic Equipments and Facilities*, (3 Volumes), Report No. FAA-RD-75-215, Contract DT-FA72WA-2850, Engineering Experiment Station, Georgia Institute of Technology, Atlanta, GA, Dec. 1975, AD A022 332, AD A022 608, and AD A022-871.

3. *Final Report on the Development of Bonding and Grounding Criteria for John F. Kennedy Space Center*, WDL-TR-4202, (3 Volumes), Contract NAS10-6879, Philco-Ford Corp., Palo Alto, CA, 30 June 1970.

4. Corey, L.E., *Engineering for Electromagnetic Interference/Electromagnetic Pulse Protection in NORAD Chyenne Mountain Complex*, ADC Communications and Electronics Digest, ADCRP-100-1, Vol. 22, No. 4, April 1972, pp. 24-30.

Index

A

Adhesive, Conductive...8.5,8.23
Amplifier Grounding..7.16,7.20
Antenna Effects................2.16,2.21-2.23,4.8,6.9,6.12,6.18,7.10

B

Balancing.........................2.23,3.5,3.6,7.11,7.15,7.18 -7.20
Battery Grounding.....................6.6,6.7,6.8,6.9,6.10,6.13,7.21
Bolts.......................................8.5,8.16,8.17,8.18,8.19
Bonding...3.1,8.1-8.25
Bond Protection Code..8.12,8.14
Bond Resistance...8.3,9.2,9.7,9.10
Bonds, Direct..8.4
Bond Straps......................8.6-8.10,8.14,8.15,8.19,8.20,8.25
Branch Circuit..5.4
Brazing...8.5,8.16

C

Capacitance.........................2.5,2.14,2.15,2.16,3.9,4.6,4.8
Capacitive Grounding...3.9
Central System...6.14,6.15
Circuit Grounding...7.2-7.24
Clustered System..6.8,6.12
Common Mode..3.5,3.6,5.5,7.2
Conductive,
    Adhesive..8.5,8.23
    Coupling..2.19,2.20,6.17,7.2
Corrosion...8.10,8.12,8.17,8.23,8.25

Index

D

Digital Circuit Ground...7.10,7.14
Distributed Systems..6.12,6.13

E

EMP...4.3,4.5,6.16
Environment,
  Noise...1.1,4.1,4.3,6.1,6.16,7.15,9.3,9.11

F

Facility Ground Network..6.1
Floating Ground..4.2
Frequency Translation..3.7
Frequency Selective Grounding..3.8

G

Galvanic Series...8.11
Green Safety Wire...2.5,5.4,7.14,7.21,9.8
Ground,
  Bus...3.3,4.5,7.15,9.7
  Digital Circuit...7.10,7.14
  Floating...4.2
  Loops...3.5,4.8,4.14
  Multi-point...3.4,4.8,4.13
  Single-point........3.5,4.3,4.7,6.5,6.16-6.19,7.6,7.7,7.11,7.15,9.12
Grounding,
  Amplifier..7.16,7.20
  Battery..6.6,6.7,6.8,6.9,6.10,6.13,7.21
  Capacitive...3.9
  Circuits...7.2,7.24
  Frequency, Selective...3.8
  Inductive..3.9
  Instrumentation..7.15,7.21
  Low Frequency..........................4.15,6.17,6.18,7.3,7.21,7.25
  Neutral...5.4,5.8,7.14,9.8,9.9
  Power Supply..............................7.4,7.5,7.8,7.9,7.10,7.13
  RF.........................3.4,4.11,4.12,4.15,6.5,6.12,7.22-7.25
  Safety............1.3,4.12,4.14,4.15,5.1-5.8,6.7,6.8,7.21,8.16,8.17
  Zonal..4.5,4.6

Index

I

Impedance..2.10,2.12,2.13,2.19
Inductance........................2.5,2.7,2.8,2.9,3.9,4.8,8.6-8.8
Inductive Grounding...3.9
Interference. def. of...2.1
Instrumentation Grounding..................................7.15,7.21
Isolated System...6.6,6.8

L

Lightning..............1.3,4.13,4.14,4.15,6.3,6.7,6.8,7.13,7.14,8.16
Low Frequency Grounding...................4.15,6.17,6.18,7.3,7.21,7.25

M

Maintenance...9.7,9.14
Multi-point Ground.......................................3.4,4.8,4.13

N

National Electrical Code..............4.15,5.3,5.5,6.3,6.9,7.14,7.21
National Lightning Protection Code................................6.3
Neutral, Grounding of........................5.4,5.8,7.14,9.8,9.9
Noise,.............
  Coupling........2.1,2.19,2.24,4.8,4.14,5.5,5.7,5.8,6.16,7.3,7.4,7.8
  Environment....................1.1,4.1,4.3,6.1,6.16,7.15,9.3,9.11

O

Optical Isolation...3.7

P

Power Supply Grounding.....................7.4,7.5,7.8,7.9,7.10,7.13

Index

R

Resistance..2.5,2.6,8.3,9.2
Resonance Effects...............................2.9,2.11,2.14-2.17,8.9
RF Grounding....................3.4,4.11,4.12,4.15,6.5,6.12,7.22-7.25

S

Safety Grounding....1.3,4.12,4.14,4.15,5.1-5.8,6.7,6.8,7.21,8.16,8.17
Shielding...2.21-2.23,8.20
Shields, grounding.....1.3,4.12,6.5,7.13,7.15,7.22,7.25,7.27,8.1,8.22
Signal Reference Plane...1.4,2.3,2.19,3.4,4.1,4.6,4.10,4.14,4.16,9.11
Shock Hazards...5.1,5.2
Single Phase..5.4
Single Point Ground..3.5,4.3,4.7,6.5,6.16-6.19,7.6,7.7,7.11,7.15,9.12
Solder..8.5,8.16,8.19
Star Ground (see single point ground)

T

Transducers...7.15,7.20
Transmission Lines..2.15,2.16
Tree Ground (see single point ground)

U

Underwriters Laboratory (UL)......................................6.3,7.14

W

Welding..8.4,8.16,8.19

Z

Zonal Grounding...4.5,4.6